Universitext

Ian Chiswell

A Course in Formal Languages, Automata and Groups

 Springer

Ian Chiswell
Department of Pure Mathematics
School of Mathematical Sciences
Queen Mary, University of London
London E1 4NS, UK
i.m.chiswell@qmul.ac.uk

ISBN 978-1-84800-939-4   ISBN 978-1-84800-940-0 (eBook)
DOI 10.1007/978-1-84800-940-0

British Library Cataloguing in Publication Data
A catalogue record for this book is available from the British Library

Library of Congress Control Number: 2008939035

Mathematics Subject Classification (2000): 03D10, 03D20, 20F10, 20F65, 68Q05, 68Q42, 68Q45

Hopcroft/Ullman, *Formal Languages and Their Relation to Automata* (adapted material from Chapter 5 (Section 4.2, Section 4.3, Theorem 5.1, Theorem 5.2, and Theorem 5.3) and Chapter 12 (Theorem 12.4 and Theorem 12.9)), © 1969. Reproduced by permission of Pearson Education, Inc.

Printed on acid-free paper

Springer Science+Business Media
springer.com

Preface

This book is based on notes for a master's course given at Queen Mary, University of London, in the 1998/9 session. Such courses in London are quite short, and the course consisted essentially of the material in the first three chapters, together with a two-hour lecture on connections with group theory. Chapter 5 is a considerably expanded version of this.

For the course, the main sources were the books by Hopcroft and Ullman ([20]), by Cohen ([4]), and by Epstein et al. ([7]). Some use was also made of a later book by Hopcroft and Ullman ([21]). The ulterior motive in the first three chapters is to give a rigorous proof that various notions of recursively enumerable language are equivalent. Three such notions are considered. These are: generated by a type 0 grammar, recognised by a Turing machine (deterministic or not) and defined by means of a Gödel numbering, having defined "recursively enumerable" for sets of natural numbers. It is hoped that this has been achieved without too many arguments using complicated notation. This is a problem with the entire subject, and it is important to understand the idea of the proof, which is often quite simple. Two particular places that are heavy going are the proof at the end of Chapter 1 that a language recognised by a Turing machine is type 0, and the proof in Chapter 2 that a Turing machine computable function is partial recursive.

Chapter 1 begins by discussing grammars and the Chomsky hierarchy, then the notion of machine recognition. It is shown that the class of regular languages coincides with the class recognised by a finite state automaton, whether or not we restrict to deterministic machines, and whether or not blank squares are allowed on the tape. There is also a discussion of Turing machines and the languages they recognise, including the result mentioned above, that a language recognised by a Turing machine is type 0. There are also further characterisations of regular languages, including Kleene's theorem that they are precisely the rational languages. The chapter ends with a brief discussion of machine recognition of context-sensitive languages, which was not included in the course.

Chapter 2 is about computable functions, and begins with a standard discussion of primitive recursive, recursive and partial recursive functions, and of primitive recursive and recursive predicates. Then various precise notions of computability are considered. These are: computation by register programs, by abacus machines and by Turing machines. In all cases, it is shown that the computable functions are precisely the partial recursive functions. The account follows [4], except that modular machines are not used. This entails giving a direct proof that Turing machine computable implies partial recursive. As mentioned above, this is heavy going, although briefer than if the theory of modular machines had been developed. To ease matters, the proof of a technical lemma has been placed in an appendix.

Chapter 3 begins with an account of recursively enumerable sets of natural numbers. Recursively enumerable languages are defined by means of Gödel numberings, and we then proceed to the proof of the main result, previously mentioned, characterising recursively enumerable languages. The comments on complexity at the end of the chapter were not included in the course, and are intended for use in Chapter 5.

Chapter 4 is about context-free languages and is material not included in the course. It is considerably heavier going than the previous three chapters. Much of the material follows the books of Hopcroft and Ullman, including their more recent one with Motwani ([22]). Some of the results are needed in Chapter 5. However, the ulterior motive for this chapter is to clarify the relationship between LR(k) languages and deterministic (context-free) languages. Neither [20] nor [21] seems to give a complete account of this.

Chapter 5 is on connections with group theory, which is a subject of great interest to the author, and a primary motivation for studying formal language theory. It begins with the author's philosophical musings on the idea of a group presentation, which are quite elementary. There is a brief discussion of free groups, free products and HNN-extensions. Most of the rest of the chapter is devoted to the word problem for groups. We prove Anisimov's theorem that a group has regular word problem if, and only if, it is finite. The highlight is a reasonably self-contained account of the result of Muller and Schupp. This says that a group has context-free word problem if and only if it is free by finite. It makes use of Dunwoody's result that a finitely presented group is accessible. To give a proof of this would have been too great a digression. A discussion of groups with word problem in other language classes is also given. The chapter ends with a brief discussion of (synchronous) automatic groups, including a proof of the characterisation by means of the fellow traveller property.

Expanding the lectures has given Chapter 5 a theme, which is the interplay between group theory, geometry (specifically, the Cayley graph) and formal language theory. It seems likely that there is a lot more to be said on this subject.

The proofs of several results have been placed in Appendix A, usually to improve the flow of the main text. In some cases, these were given as handouts to the class. Appendices B and C were also handouts, although Appendix B has been expanded to include a brief discussion of universal Turing machines. Appendix D contains solutions to selected exercises. A complete solutions manual, password protected, is available to instructors via the Springer website. To apply for a password, visit the book webpage at *www.springer.com* or email *textbooks@springer.com*. The number of exercises is fairly small, and they vary in difficulty; some of them can be used as templates for similar exercises (only the exercises in Chapters 1 and 2 were actually used in the course).

The impetus for the development of formal language theory comes from computer science, and as already noted, it can be at times quite complicated. Despite this, it is an elegant part of pure mathematics. The book is written by a mathematician and intended for mathematicians. Nevertheless, it is hoped it may be of some interest to computer scientists.

The book can be viewed only as an introduction to the subject (the audience consisted of graduate students in mathematics). For further reading on formal languages, see, for example, [33] and [34].

The prerequisite for understanding the book is some exposure to abstract mathematics, including an understanding of some basic ideas, such as mapping, Cartesian product and equivalence relation (note that "mapping" and "function" mean the same thing throughout the book). At various points the reader is assumed to be familiar with the combinatorial idea of a graph. This includes both directed and undirected graphs and the idea of a tree. Generally, vertices of a graph are denoted by circles or dots, but in the case of parsing trees (Chapter 4) they are indicated only by their labels. Of course, in Chapter 5, some knowledge of basic group theory is assumed. Also, the reader needs to know at least the definition of a semigroup and a monoid. No advanced mathematical knowledge is needed.

Concerning notation, words in a formal language are elements of a Cartesian product A^n, where n is an integer, and in this context are usually written without commas and parentheses. In other cases where Cartesian products are involved, for example the transitions of a machine or the definition of grammars and machines, commas and parentheses are used. The exception is in writing the transitions of a Turing machine, in order to conform with what appears to be the usual practice. Our definitions of grammars and machines are quite formal. This seems the best way to proceed, although it has gone out of fashion when defining basic mathematical objects (such as a group). As usual, $\mathbb{R}$ denotes the set of real numbers, $\mathbb{Q}$ the set of rational numbers, $\mathbb{Z}$ the set of integers and $\mathbb{N}$ the set of natural numbers, which in this book means $\{0, 1, 2, \ldots\}$.

The author thanks Sarah Rees, Claas Röver and Richard Thomas for their helpful conversations and email messages. In particular, several of the arguments in Chapter 5 were suggested by Richard Thomas. He also warmly thanks Daniel Cohen for his very useful and perceptive comments on the manuscript.

A list of errata will be available on the book webpage at *www.springer.com*.

Queen Mary, University of London   *Ian Chiswell*
School of Mathematical Sciences
June 2008

Contents

Chapter 1
Grammars and Machine Recognition

By a language we have in mind a written language. Such a language, whether natural or a programming language, has an *alphabet*, and words are formed by writing strings of letters in the alphabet. (In the case of some natural languages, the alphabet for this purpose may not be what is normally described as the alphabet.) However, to develop a mathematical theory, we need precise definitions of these ideas. An alphabet consists of a number of letters, which are written in a certain way. However, the letters are not physical entities, but abstract concepts. If one writes "a" twice, the two copies will not look identical, but one hopes they are sufficiently close to be recognised as representing the abstract concept of the first letter of the alphabet.

To the pure mathematician, this presents no problem. An alphabet is just a set. The words are then just finite sequences of elements of the alphabet. Allowing such a wide-ranging definition will turn out to be very convenient. The alphabet can even be infinite, although in this book it is usually finite. (The exceptions are in the definition of abacus machines in Chap. 2, and the discussion of free products and HNN-extensions in Chap. 5.)

Thus, let A be a set and let A^m be the set of all finite sequences $a_1 \ldots a_m$ with $a_i \in A$ for $1 \leq i \leq m$. Elements of A are called *letters* or *symbols*, and elements of A^m are called *words* or *strings* over A of *length m*.

Note: m is a natural number; $A^0 = \{\varepsilon\}$, where ε is the *empty word* having no letters, and A^1 can be identified with A. The set A^m $(m \geq 2)$ can be identified with the Cartesian product $\underbrace{A \times A \times \ldots \times A}_{m \text{ copies}}$, but its elements are written without the usual commas and parentheses.

Definition. Put $A^+ = \bigcup_{m \geq 1} A^m$, $A^* = \bigcup_{m \geq 0} A^m = A^+ \cup \{\varepsilon\}$.

If $\alpha = a_1 \ldots a_m$, $\beta = b_1 \ldots b_n \in A^*$, define $\alpha\beta$ to be $a_1 \ldots a_m b_1 \ldots b_n$ (an element of A^{m+n}). This gives a binary operation on A^* (and on A^+) called *concatenation*. It is associative: $\alpha(\beta\gamma) = (\alpha\beta)\gamma$ and $\alpha\varepsilon = \varepsilon\alpha = \alpha$. Thus A^+ is a semigroup (the *free semigroup on A*) and A^* is a monoid (the *free monoid on A*). Denote the length of a word α by $|\alpha|$. As usual, we can define α^n, where $n \in \mathbb{N}$, by: $\alpha^0 = \varepsilon$, $\alpha^{n+1} = \alpha^n \alpha$.

I. Chiswell, *A Course in Formal Languages, Automata and Groups*,
DOI 10.1007/978-1-84800-940-0_1,
© Springer-Verlag London Limited 2009

If α is a word over an alphabet A, a *subword* of α is a word $\gamma \in A^*$ such that $\alpha = \beta\gamma\delta$ for some β, $\delta \in A^*$. If $\alpha = \beta\gamma$, then β is called a *prefix* of α and γ is called a *suffix* of α.

Definition. A *language with alphabet A* is a subset of A^*.

We shall consider languages defined in a particular way, using what is called a rewriting system. This is essentially a set of rules, each of which allows some string u, whenever it occurs in a word, to be replaced by another string v. Such a rule is specified by the ordered pair (u, v), leading to the following formal definitions.

Definition. A *rewriting system* on A is a subset of $A^* \times A^*$.

If R is a rewriting system and $(\alpha, \beta) \in R$, then for any u, $v \in A^*$, we say that $u\alpha v$ *rewrites* to $u\beta v$. Elements of R are written as $\alpha \longrightarrow \beta$ rather than (α, β).

Definition. For $u, v \in A^*$, $u \overset{\bullet}{\longrightarrow} v$ means there is a finite sequence $u = u_1, \ldots, u_n = v$ of elements of A^* such that u_i rewrites to u_{i+1} for $1 \le i \le n - 1$. Such a sequence is called an R-derivation of v from u. (Write $u \overset{\bullet}{\underset{R}{\longrightarrow}} v$ if necessary.)

Definition. A *grammar* is a quadruple (V_N, V_T, P, S) where

(1) V_N, V_T are disjoint finite sets (the set of *non-terminal* and *terminal* symbols respectively).
(2) $S \in V_N$ (the *start symbol*).
(3) P is a finite rewriting system on $V_N \cup V_T$.

(Elements of P are called *productions* in this context.)

Definition. The language L_G generated by G is

$$L_G = \left\{ w \in V_T^* \mid S \overset{\bullet}{\longrightarrow} w \right\}$$

(a language with alphabet V_T).

Definition. A production is *context-free* if it has the form $A \longrightarrow \alpha$, where $A \in V_N$ and $\alpha \in (V_N \cup V_T)^+$. It is *context-sensitive* if it has the form $\beta A \gamma \longrightarrow \beta \alpha \gamma$, where $A \in V_N$, $\alpha, \beta, \gamma \in (V_N \cup V_T)^*$, $\alpha \neq \varepsilon$.

The reason for the names is that in using a context-free production $A \longrightarrow \alpha$ in a derivation, A can be replaced in a word by the word α regardless of the context (the strings of letters that appear to the left and right of A in the word). With a context-sensitive production $\beta A \gamma \longrightarrow \beta \alpha \gamma$, whether or not it can be used to replace A by γ depends on the context (β must occur to the left, and γ to the right of A in the word). Note, however, that $\beta = \gamma = \varepsilon$ is allowed in the definition of context-sensitive production, so context-free productions are context-sensitive.

The Chomsky hierarchy. This is a sequence of four classes of grammars (and corresponding classes of languages), each contained in the next.

A grammar G as defined above is said to be of *type 0*. It is of *type 1* if all productions have the form $\alpha \longrightarrow \beta$ with $|\alpha| \le |\beta|$.

Note. It can be shown that, if G is of type 1, then $L_G = L_{G'}$ for some context-sensitive grammar G', that is, a grammar in which all productions are context-sensitive, in the sense above. See Lemma A.2 in Appendix A.

A grammar G is of *type* 2 (or *context-free*) if all productions are context-free. It is of *type* 3 (or *regular*) if all productions have the form $A \longrightarrow aB$ or $A \longrightarrow a$, where A, $B \in V_N$ and $a \in V_T$.

A language L is of type n if $L = L_G$ for some grammar G of type n ($0 \le n \le 3$). We also use regular, context-free and context-sensitive to describe languages of types 3, 2 and 1, respectively.

The idea of a context-free grammar was introduced by Chomsky as a possible way of describing natural languages. Although they have not proved successful in this, context-free languages have turned out to be important in describing programming languages. The first such descriptions were for FORTRAN by Backus [1], and ALGOL by Naur [28]. Indeed, context-free grammars are sometimes called Backus-Naur form grammars. For an example of a modern language (HTML) described by a context-free language, see [22, §5.3.3].

Context-free languages are important in the design of compilers, in particular the design of parsers. For a discussion of the uses of context-free languages, we refer to [22, §5.3].

Examples. It is left to the reader to prove that L_G is as claimed. This is easy in Examples (1)-(4); Example (5) is discussed in [20, Example 2.2].

(1) Let $G = (\{S\}, \{0\}, P, S)$ where P consists of

$$S \longrightarrow 0, \quad S \longrightarrow 0S.$$

Then $L_G = \{0^n \mid n \ge 1\} = \{0\}^+$ (a type 3 language).

(2) Let $G = (\{S, A\}, \{0, 1\}, P, S)$ where P consists of

$$S \longrightarrow 0S, \quad S \longrightarrow A, \quad A \longrightarrow 1A, \quad A \longrightarrow 1.$$

Then $L_G = \{0^m 1^n \mid m \ge 0, \, n \ge 1\}$ (also type 3).

(3) Let $G = (\{S\}, \{0, 1\}, P, S)$ where P consists of

$$S \longrightarrow 0S1, \quad S \longrightarrow 01.$$

Then $L_G = \{0^n 1^n \mid n \ge 1\}$ (a type 2 language).

(4) Let $G = (\{S, A\}, \{a, b, c\}, P, S)$ where P contains

$$S \longrightarrow Sc, \quad S \longrightarrow A, \quad A \longrightarrow aAb, \quad A \longrightarrow ab.$$

Then $L_G = \{a^n b^n c^i \mid n \ge 1, \, i \ge 0\}$ (also type 2).

(5) Let $G = (\{S, B, C\}, \{a, b, c\}, P, S)$ where P is

$$S \longrightarrow aSBC$$
$$S \longrightarrow aBC$$
$$CB \longrightarrow BC$$
$$aB \longrightarrow ab$$
$$bB \longrightarrow bb$$
$$bC \longrightarrow bc$$
$$cC \longrightarrow cc$$

Then $L_G = \{a^n b^n c^n \mid n \geq 1\}$ (type 1).

The Empty Word. If L is a type n language ($1 \leq n \leq 3$), it is easy to see that $\varepsilon \notin L$. However, it is useful to view $L \cup \{\varepsilon\}$ as also a language of type n. To do this, we make the following convention:

$S \longrightarrow \varepsilon$ is allowed as a production for type n grammars ($1 \leq n \leq 3$), provided S does not occur on the right-hand side of any production.

To see that this works, we need to prove the following lemma.

Lemma 1.1. *If L is a type n language ($1 \leq n \leq 3$), then $L = L_{G_1}$ for some grammar G_1 of type n, whose start symbol S_1 does not occur on the right-hand side of any production of G_1.*

Proof. Let $L = L_G$, where $G = (V_N, V_T, P, S)$ is a type n grammar. Let S_1 be a letter not in $V_N \cup V_T$ and put $G_1 = (V_N \cup \{S_1\}, V_T, P_1, S_1)$, where

$$P_1 = P \cup \{S_1 \longrightarrow \alpha \mid S \longrightarrow \alpha \text{ is in } P\}.$$

Then G_1 is of type n and S_1 does not occur on the right-hand side of any production of G_1.

Suppose $S \xrightarrow[P]{\bullet} w$, so there is a P-derivation $S = u_1, \ldots, u_n = w$, so $S \longrightarrow u_2$ is in P, hence $S_1 \longrightarrow u_2$ is in P_1; also, $P \subseteq P_1$, so $S_1, u_2, \ldots, u_n = w$ is a P_1-derivation. Hence $S_1 \xrightarrow[P_1]{\bullet} w$.

Conversely, suppose $S_1 \xrightarrow[P_1]{\bullet} w$ and let $S_1 = u_1, u_2, \ldots, u_n = w$ be a P_1-derivation. Then $S_1 \longrightarrow u_2$ is in P_1, so $S \longrightarrow u_2$ is in P, and S_1 does not occur in $u_2, \ldots, u_n$ since it does not occur in the right-hand side of a production in P_1. Hence $S, u_2, \ldots, u_n$ is a P-derivation, so $S \xrightarrow[P]{\bullet} w$. Thus $L = L_{G_1}$.   □

We can now show that our convention works.

Corollary 1.2. *If L is of type n ($1 \leq n \leq 3$), then $L \cup \{\varepsilon\}$ and $L \setminus \{\varepsilon\}$ are of type n.*

Proof. By Lemma 1.1, $L = L_G$ where G is some grammar of type n whose start symbol S does not occur on the right-hand side of any production of G. Adding $S \longrightarrow \varepsilon$ to the set of productions gives a grammar of type n generating $L \cup \{\varepsilon\}$, since the only derivation using $S \longrightarrow \varepsilon$ is S, ε. If $\varepsilon \in L_G$, the set P of productions must contain $S \longrightarrow \varepsilon$. Removing this from P gives a type n grammar generating $L \setminus \{\varepsilon\}$.
                                            □

Machine Recognition. We consider imaginary machines which have (at least) a read head which can read a tape. The tape is divided into squares on which are written letters from an alphabet, and the head can scan a single square. Depending on the letter scanned and other things, the machine can move to an adjacent square and in some cases, alter the letter scanned. When started with a string on the tape, the machine either accepts or rejects the string in some manner. The language recognised by the machine is the set of strings which it accepts.

Associated to each type in the Chomsky hierarchy is a class of machines, such that a language is recognised by a machine in the class if and only if it is defined by a grammar of the appropriate type. The classes of machines involved are listed in the following table.

Language type	Recognised by a
0	Turing machine
1	linear bounded automaton
2	non-deterministic pushdown stack automaton
3	finite state automaton

In this chapter we shall only look at the machines involved with type 0 and type 3 grammars, beginning with type 3.

Finite State Automata

Definition. A finite state automaton (which will always be abbreviated to FSA) is a quintuple $M = (Q, F, A, \tau, q_0)$, where

(1) Q is a finite set (the set of *states*).
(2) F is a subset of Q (the set of *final* states).
(3) A is a finite set (the *alphabet*).
(4) $\tau \subseteq Q \times A \times Q$ (the set of *transitions*).
(5) $q_0 \in Q$ (the *initial state*).

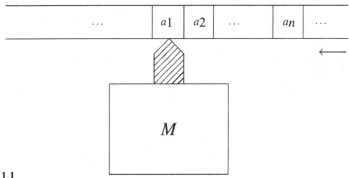

Figure 1.1

The idea is that M has a read head scanning a tape divided into squares, each of which has a symbol from A printed on it. There is no restriction on the length of the tape. Further, M scans one square at a time, and is in one of a finite number of states (represented by the elements of Q). If M is in state q, reading a on the tape, and $(q, a, q') \in \tau$, then M can change to state q' and move the tape one square to the left (equivalently, move the head one square to the right).

Definition. A *computation* of M is a sequence $q_0, a_1, q_1, a_2, q_2, \ldots, a_n, q_n$ (with $n \geq 0$) where $(q_{i-1}, a_i, q_i) \in \tau$ for $1 \leq i \leq n$.
The *label* on the computation is $a_1 \ldots a_n$. The computation is *successful* if $q_n \in F$.

(The idea is that M successively reads $a_1, \ldots, a_n$ on the tape, passing through the states $q_0, q_1, \ldots, q_n$ as it does so.)
A string $a_1 \ldots a_n$ is *accepted* by M if there is a successful computation with label $a_1 \ldots a_n$.

Definition. The language recognised by M is

$$L(M) = \{w \in A^* \mid w \text{ is accepted by } M\}.$$

Transition Diagram. The transition diagram of a FSA is a directed graph, with vertex set Q, the set of states, and an edge for each transition. The edge corresponding to $(q, a, q') \in \tau$ runs from q to q', and has label a. Also, some vertices are labelled; the initial state q_0 is labelled with "$-$" and every final state is labelled with "$+$". It is drawn by enlarging the circles representing the vertices and writing their labels inside the circles.

Note that there is a one-to-one correspondence

$$\text{computations of } M \longleftrightarrow \text{paths in the graph starting at } q_0$$

(If $q_0, e_1, q_1, \ldots, e_n, q_n$ is a path, replace each edge e_i by its label to get the corresponding computation.)

A FSA can be specified by its transition diagram. A finite directed graph with edge labels from a set A is the transition diagram of a FSA provided: if q, q' are vertices and $a \in A$, no more than one edge from q to q' has label a, exactly one vertex is labelled "$-$" and some (possibly no) vertices are labelled "$+$".

Note. The label on the computation q_0 is ε, so $\varepsilon \in L(M)$ if and only if $q_0 \in F$. In this case, $\pm$ is drawn in the circle representing q_0.

Examples. In these examples, it is left to the reader to show that the language recognised is as claimed.

(1) Let the alphabet A have a single letter, say $A = \{a\}$, and let the transition diagram be

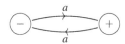

Figure 1.2

If M is the corresponding FSA, $L(M) = \{a^{2n+1} \mid n = 0, 1, 2, \ldots\}$.

(2) Let $A = \{a, b\}$, with M the FSA defined by the transition diagram

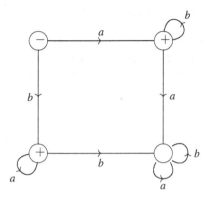

Figure 1.3

Then $L(M) = \{ab^n \mid n \in \mathbb{N}\} \cup \{ba^n \mid n \in \mathbb{N}\}$.

(3) Let A be any finite set, with the transition diagram having no edges and one vertex, which is a final state:

then $L(M) = \{\varepsilon\}$.

(4) Again let A be any finite set, and suppose $a_1, \ldots, a_n \in A$, with transition diagram

Then $L(M) = \{a_1 \ldots a_n\}$. Note that (3) can be viewed as the case $n = 0$ of (4).

It is possible to have two transitions (q, a, q') and (q, a, q'') with $q' \neq q''$ (more than one edge leaving q with the same label). Then if the FSA is in state q reading a, it can enter either state q' or state q'' (or possibly other states). This is not a problem with our definition of a computation, although if we imagine a machine actually running a computation, it would need some means of deciding which state to move to.

Also, given state q and a in the alphabet A, there may be no transition of the form (q, a, q'), so if the FSA is in state q reading a, it grinds to a halt.

Definition. A FSA is *deterministic* if for all $q \in Q$, $a \in A$, there is exactly one $q' \in Q$ such that $(q, a, q') \in \tau$.

If M is deterministic, we denote the unique such q' by $\delta(q, a)$, thereby defining a function $\delta : Q \times A \to Q$, called the *transition function* of M.

We extend δ to a mapping $Q \times A^* \to Q$ recursively by

$$\delta(q, \varepsilon) = q$$
$$\delta(q, wa) = \delta(\delta(q, w), a) \quad \text{for } w \in A^*, \, a \in A.$$

The idea is that if M is in state q and successively reads the letters of w on the tape, it will be in state $\delta(q, w)$. Thus if $q_0, a_1, q_1, a_2, \ldots, a_n, q_n$ is a computation, then $q_n = \delta(q_0, a_1 \ldots a_n)$ (this is easily proved by induction on n). Consequently, $L(M) = \{w \in A^* \mid \delta(q_0, w) \in F\}$.

In defining $\delta(q, \varepsilon) = q$, we are making the convention that, if the tape is blank, M does not change state. However, we can take a different point of view, that blank squares are allowed even if the tape is not blank, and M can change state when a blank square is read.

Definition. A *generalised* FSA M is one in which triples of the form (q, ε, q') are allowed as transitions (so $\tau \subseteq Q \times (A \cup \{\varepsilon\}) \times Q$ and ε is allowed as a label on edges of the transition diagram). The language $L(M)$ is defined as before. (Note, however, that if $a_i = \varepsilon$, $a_1 \ldots a_n = a_1 \ldots a_{i-1} a_{i+1} \ldots a_n$.)

We show that, whatever notion of FSA is used, the class of languages recognised is the same.

Proposition 1.3. *Let L be a language with alphabet A. The following are equivalent:*

(1) *L is recognised by a deterministic FSA.*
(2) *L is recognised by a FSA.*
(3) *L is recognised by a generalised FSA.*

Proof. Clearly $(1) \Rightarrow (2) \Rightarrow (3)$, and we show $(3) \Rightarrow (1)$. Suppose L is recognised by a generalised FSA $M = (Q, F, A, \tau, q_0)$. If $X \subseteq Q$, let $\overline{X}$ be the set of all possible endpoints of paths in the transition diagram for M starting at a vertex of X and such that the label on all edges of the path is ε. Note that $X \subseteq \overline{X}$ and $\overline{\overline{X}} = \overline{X}$.

Define a deterministic FSA $M' = (Q', F', A, \tau', q_0')$ as follows. Put $Q' = $ the set of all subsets X of Q such that $X = \overline{X}$

$$\delta(X, a) = \overline{\{\text{all endpoints of edges labelled } a \text{ which start at a vertex of } X\}}$$

(so $\tau' = \{(X, a, \delta(X, a)) \mid X \in Q', \, a \in A\}$),
$q_0' = \overline{\{q_0\}}$ and $F' = \{X \in Q' \mid q \in X \text{ for some } q \in F\}$.

It is left as an exercise to show that $L(M') = L(M) = L$. □

Definition. If (1)-(3) in Prop. 1.3 are satisfied, we say that L is recognised by a FSA.

We can now establish the relationship between FSA's and regular languages mentioned previously.

Theorem 1.4. *For a language L, the following are equivalent:*

(1) *L is a type 3 (regular) language.*
(2) *L is recognised by a FSA.*

Proof. (1) $\Rightarrow$ (2). Let $L = L_G$ where $G = (V_N, V_T, P, S)$ is a type 3 grammar. Define a FSA $M = (V_N \cup \{X\}, F, V_T, \tau, S)$ (where X is a new letter not in $V_N \cup V_T$) by

$$F = \begin{cases} \{S, X\} & \text{if } S \longrightarrow \varepsilon \text{ is in } P \\ \{X\} & \text{otherwise} \end{cases}$$

and $\tau = \{(B, a, C) \mid B \longrightarrow aC \text{ is in } P\} \cup \{(B, a, X) \mid B \longrightarrow a \text{ is in } P \text{ (and } a \neq \varepsilon)\}$.

We show $L_G = L(M)$. Suppose $u = a_1 \ldots a_n \in L_G$ $(n \geq 1)$, so there is a P-derivation

$$S, a_1 A_1, a_1 a_2 A_2, \ldots, a_1 \ldots a_{n-1} A_{n-1}, a_1 \ldots a_n$$

Then $(S, a_1, A_1), (A_1, a_2, A_2), \ldots, (A_{n-2}, a_{n-1}, A_{n-1}), (A_{n-1}, a_n, X)$ are in τ, so

$$S, a_1, A_1, a_2, A_2, \ldots A_{n-1}, a_n, X$$

is a successful computation of M, hence $u \in L(M)$. If $\varepsilon \in L_G$ then $S \in F$, so $\varepsilon \in L(M)$. Thus $L_G \subseteq L(M)$.

To show the reverse inclusion, suppose $u = a_1 \ldots a_n \in L(M)$ $(n \geq 1)$, so there is a computation

$$S, a_1, A_1, a_2, A_2, \ldots A_{n-1}, a_n, X$$

of M (if $S \longrightarrow \varepsilon \in P$, S does not appear on the right-hand side of any production, so it can't end with A_{n-1}, a_n, S). Then P contains

$$S \longrightarrow a_1 A_1, \ldots, A_{n-2} \longrightarrow a_{n-1} A_{n-1}, A_{n-1} \longrightarrow a_n$$

(because X does not occur in any production). Hence $S \stackrel{\bullet}{\longrightarrow} a_1 \ldots a_n = u$. If $\varepsilon \in L(M)$ then $S \in F$, so $S \longrightarrow \varepsilon \in P$, hence $\varepsilon \in L_G$. Thus $L_G = L(M)$.

(2) $\Rightarrow$ (1). Suppose $L = L(M)$ where $M = (Q, F, A, \tau, q_0)$ is a deterministic FSA. We can assume $Q \cap A = \emptyset$. Put $G = (Q, A, P, q_0)$, where

$$P = \{B \longrightarrow aC \mid (B, a, C) \in \tau\} \cup \{B \longrightarrow a \mid (B, a, C) \in \tau \text{ and } C \in F\}.$$

Then for $u \in A^*$, $u \neq \varepsilon$, $S \stackrel{\bullet}{\longrightarrow} u$ if and only if $u \in L(M)$, by a similar argument, left to the reader. If $q_0 \in F$, then $\varepsilon \in L(M)$ and $L(M) = L_G \cup \{\varepsilon\}$, otherwise $L(M) = L_G$. By Cor. 1.2, $L(M)$ is regular. $\qquad \square$

Remark 1.1. The alert reader will have noticed a lack of symmetry in the definition of a regular grammar, which can now be resolved. We can define a *left regular* grammar to be one in which all productions are of the form $A \longrightarrow Ba$ or $A \longrightarrow a$, where A, $B \in V_N$ and $a \in V_T$. Now if $w = a_1 \ldots a_n$ is a word in some alphabet, we define its reversal w^R to be $a_n \ldots a_1$. If L is a language, we define $L^R = \{w^R \mid w \in L\}$. If G is a regular grammar generating L, and all productions $A \longrightarrow aB$ are replaced by

$A \longrightarrow Ba$, we obtain a left regular grammar generating L^R. Similarly, if a left regular grammar generates a language L, we obtain a regular grammar generating L^R.

We claim that a language L is regular if and only if L^R is. Since $(L^R)^R = L$, it follows that a language is regular if and only if it is generated by a left regular grammar. Again since $(L^R)^R = L$, it suffices to show that, if L is regular, then so is L^R. To see this, take a FSA recognising L, and modify its transition diagram as follows. Reverse the direction of all edges and make the initial state the only final state. Add a new initial state, and add an edge from it to each of the original final states, with label ε. This is the transition diagram of a generalised FSA recognising L^R. (See also Remark 4.4.)

One can also ask what happens if productions of the form $A \longrightarrow aB$ and $A \longrightarrow Ba$ are both allowed. This leads to a class known as *linear* languages (see Exercises 4–6 in Chapter 4).

Rational Operations on Languages. Let L, L_1, L_2 be languages with alphabet A. The following are also languages with alphabet A.

(1) L^*; strictly, this is a language with alphabet L, but a finite sequence $u_1 \ldots u_m$, $u_i \in L$, can be viewed as the concatenation of the words $u_1, \ldots, u_m$, so an element of A^*. (Algebraically, L^* is the submonoid of A^* generated by L.)
(2) $L_1 L_2 = \{uv \mid u \in L_1, \ v \in L_2\}$.
(3) $L_1 \cup L_2, L_1 \cap L_2$ and $L^c = A^* \setminus L$.

The language $L_1 L_2$ is called the *product* of L_1 and L_2, and the operation which associates L^* to L is called *Kleene star*.

Lemma 1.5. *Let L, L_1 and L_2 be languages.*

(1) *If L is finite, it is regular.*
(2) *If L is regular then L^* is regular.*
(3) *If L_1 and L_2 are regular then $L_1 \cup L_2$ is regular.*
(4) *If L_1 and L_2 are regular then $L_1 L_2$ is regular.*
(5) *If L is regular then L^c is regular.*
(6) *If L_1 and L_2 are regular then $L_1 \cap L_2$ is regular.*

Proof. (1) From earlier examples, a language with just one word is recognised by a FSA, so is regular by Theorem 1.4. Thus (1) follows from (3).

(2) If L is regular, $L = L(M)$ for some FSA M by Theorem 1.4. Let M' be the (generalised) FSA obtained from M by making the following changes to the transition diagram.

   (i) Adding a new vertex, which is to be the only final state of M', and adding edges from each old final state of M to the new vertex, all with label ε.
  (ii) Adding another new vertex, which is to be the initial state of M', and adding an edge with label ε from the new vertex to the old initial state of M.
 (iii) Adding an edge from the new final state to the new initial state of M', and an edge in the opposite direction from the initial state to the final state, both with label ε.

This is illustrated by:

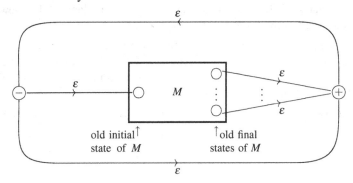

Figure 1.4

It is easy to see that $L^* = L(M')$, so L^* is recognised by a FSA, hence is regular by Theorem 1.4.

(3) By Theorem 1.4, $L_i = L(M_i)$ for $i = 1, 2$, where M_i is a FSA, and we can assume that M_1 and M_2 have no states in common. We construct a new FSA, whose transition diagram is the union of the transition diagrams for M_1 and M_2, modified as follows. There is one extra vertex as initial state and two extra edges from this new vertex to the initial states of M_1 and M_2, having label ε. The final states are those of M_1 and M_2.

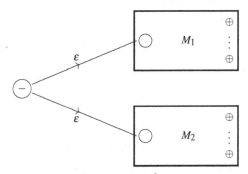

↑ old initial states of M_1 and M_2

Figure 1.5

Clearly the new FSA recognises $L_1 \cup L_2$.

(4) Let $L_i = L(M_i)$ as in the previous part. We obtain a FSA recognising $L_1 L_2$ by connecting the transition diagrams "in series", as illustrated in the diagram below.

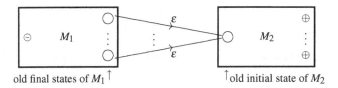

old final states of M_1 ↑   ↑ old initial state of M_2

Figure 1.6

Thus, we take the union of the transition diagrams of M_1 and M_2, with new edges from the final states of M_1 to the initial state of M_2, all with label ε. The new initial state is that of M_1, and the final states are those of M_2.

(5) If L is regular, we can write $L = L(M)$, where $M = (Q, F, A, \tau, q_0)$ is deterministic, using Prop. 1.3 and Theorem 1.4. Then $L^c = L(M')$, where $M' = (Q, Q \setminus F, A, \tau, q_0)$. (For $w \in A^*$, there is exactly one path in the transition diagram of M, starting at q_0 and with label w.)

(6) This follows from (3) and (5) by the de Morgan law: $L_1 \cap L_2 = (L_1^c \cup L_2^c)^c$. □

The *rational operations* on languages are union, product and Kleene star. Let $\mathscr{L}$ be a collection of languages with alphabet A. Call $\mathscr{L}$ *rationally closed* if, for all languages L, L_1 and L_2 with alphabet A,

(1) if L is finite, then $L \in \mathscr{L}$;
(2) if $L \in \mathscr{L}$ then $L^* \in \mathscr{L}$;
(3) if $L_1, L_2 \in \mathscr{L}$, then $L_1 \cup L_2 \in \mathscr{L}$;
(4) if $L_1, L_2 \in \mathscr{L}$, then $L_1 L_2 \in \mathscr{L}$.

There is a smallest rationally closed collection, namely the intersection of all such collections $\mathscr{L}$, which will be denoted by $\mathscr{R}$. A language L is called *rational* if $L \in \mathscr{R}$.

Theorem 1.6. (Kleene) *A language is rational if and only if it is regular.*

Proof. By Lemma 1.5, the class of regular languages on an alphabet A is rationally closed, and so contains $\mathscr{R}$. That is, a rational language is regular.

Conversely, suppose L is regular, so $L = L(M)$ for some FSA $M = (Q, F, A, \tau, q_0)$, by Theorem 1.4.

If $q, e_1, q_1, \ldots q_{n-1}, e_n, q'$ is a path in the transition diagram of M, the *intermediate states* of the path are defined to be $q_1, \ldots, q_{n-1}$. For $q, q' \in Q$, $X \subseteq Q$, let

$$L(q, q', X) = \text{the set of all labels on paths from } q \text{ to } q' \text{ for which all}$$
$$\text{intermediate states of the path belong to } X.$$

We prove by induction on the number of elements of X that $L(q, q', X) \in \mathscr{R}$.

If $X = \emptyset$, let $e_1, \ldots, e_r$ be the edges starting at q, ending at q', with labels $a_1, \ldots, a_r$ respectively. Then $L(q, q', X) = \begin{cases} \{a_1, \ldots, a_r\} & \text{if } q \neq q' \\ \{\varepsilon, a_1, \ldots, a_r\} & \text{if } q = q' \end{cases}$ is finite, so belongs to $\mathscr{R}$.

If $X \neq \emptyset$, choose $x \in X$, and define

$$L_1 = L(q, q', X \setminus \{x\}), \quad L_2 = L(q, x, X \setminus \{x\})$$
$$L_3 = L(x, x, X \setminus \{x\}) \quad L_4 = L(x, q', X \setminus \{x\}).$$

By induction, $L_i \in \mathscr{R}$ for $1 \leq i \leq 4$. Since $\mathscr{R}$ is rationally closed, $L(q, q', X) = L_1 \cup (L_2 L_3^* L_4) \in \mathscr{R}$, completing the induction. Finally, $L(M) = \bigcup_{q \in F} L(q_0, q, Q) \in \mathscr{R}$. □

Thus, although complement and intersection are not used in defining the set of rational languages, it turns out that the set of rational languages is closed under these operations, by Lemma 1.5 and Theorem 1.6.

Next, we shall prove some results on regular languages which are useful in deciding if specific languages are regular, beginning with a characterisation by certain equivalence relations.

Definition. The *index* of an equivalence relation is the number of sets in the corresponding partition.

Definition. An equivalence relation on A^* (A being any set) is *right invariant* if, for all $x, y \in A^*$, xRy implies that for all $z \in A^*$, $xzRyz$.

If L is a language with alphabet A, we can define a binary relation R_L on A^* by: xR_Ly if and only if $\chi_L(xz) = \chi_L(yz)$ for all $z \in A^*$, where χ_L is the *characteristic function* of L, that is, $\chi_L(w) = \begin{cases} 1 & \text{if } w \in L \\ 0 & \text{if } w \in A^* \setminus L \end{cases}$

Then R_L is a right-invariant equivalence relation.

Theorem 1.7. (Myhill-Nerode) *For a language with alphabet A, the following are equivalent.*

(1) *L is recognised by a FSA.*
(2) *L is the union of some of the equivalence classes of a right-invariant equivalence relation of finite index on A^*.*
(3) *R_L is of finite index.*

Proof. (1) $\Rightarrow$ (2) Suppose L is recognised by $M = (Q, F, A, \tau, q_0)$, a deterministic FSA. Let the transition function be δ. Define xRy to mean $\delta(q_0, x) = \delta(q_0, y)$, for $x, y \in A^*$. This is an equivalence relation of finite index on A^* (the index is at most the number of states of M, since $\delta(q_0, x) \in Q$). By induction on $|z|$ (where z is as in the definition of right-invariant), R is right-invariant. Finally, L is the union of those equivalence classes containing an element x such that $\delta(q_0, x) \in F$.

(2) $\Rightarrow$ (3) Let L be the union of some of the equivalence classes of R, a right-invariant equivalence relation of finite index on A^*. Then xRy implies xR_Ly. For if xRy, then $xzRyz$ for all $z \in A^*$, hence $xz \in L$ if and only if $yz \in L$, i.e. xR_Ly. Hence R_L has finite index (each R-equivalence class is contained in an R_L-equivalence class).

(3) $\Rightarrow$ (1) Assume R_L is of finite index. Let Q be the finite set of equivalence classes of R_L, and denote the equivalence class of x by $[x]$. Put $\delta([x], a) = [xa]$ for $a \in A$ (this is well-defined), $q_0 = [\varepsilon]$ and $F = \{[x] \mid x \in L\}$ to define a deterministic FSA M which recognises L (because $\delta(q_0, y) = [y]$ for $y \in A^*$, by induction on $|y|$). $\square$

Example. In Example (3), p.3, we saw that $L = \{0^n1^n \mid n > 0\}$ is type 2, but it is not type 3 (regular). Otherwise R_L has finite index, so $0^mR_L0^n$ for some $m, n > 0$ with $m \neq n$. But then $0^m1^nR_L0^n1^n$, a contradiction since $0^m1^n \notin L$ and $0^n1^n \in L$.

The next result can also be used to show L is not regular, and is another useful criterion. If v is a subword of a word $w \in L$, where L is a language, then we say that v can be "pumped" if replacing v in w by v^i, for any $i \in \mathbb{N}$, results in a word in L.

Lemma 1.8. (The Pumping Lemma) *Let L be a regular language. There is an integer $p > 0$ such that any word $x \in L$ with $|x| \geq p$ is of the form $x = uvw$, where $|v| > 0$, $|uv| \leq p$ and $uv^i w \in L$ for all $i \geq 0$.*

Proof. Let p be the number of states in a FSA recognising L, and let the FSA have initial state q_0. An accepted word $x = a_1 \ldots a_n$ is the label on a path in the transition diagram, starting at q_0 and ending at a final state, say $q_0, e_1, q_1, \ldots, e_n, q_n$. There are $n + 1$ occurrences of states in this sequence, so if $n \geq p$, there must be integers $r < s$ such that $q_r = q_s$. Choose s as small as possible subject to this. Now put $u = a_1 \ldots a_r$, $v = a_{r+1} \ldots a_s$, $w = a_{s+1} \ldots a_n$, so $|v| > 0$. The vertices $q_0, \ldots, q_{s-1}$ are distinct, by minimality of s, hence $|uv| = s \leq p$. Also, $q_r, e_{r+1}, q_{r+1}, \ldots, e_s, q_s$ is a closed path, so can be repeated $i \geq 0$ times in the original path, to give a path from q_0 to q_n with label $uv^i w$. (When $i = 0$, the path is $q_0, e_1, \ldots, q_r, e_{s+1}, q_{s+1}, \ldots, q_n$.) $\qquad\square$

There is also a pumping lemma for type 2 (context-free) languages, which is stated here to illustrate its use. Its proof is deferred until later (after Theorem 4.10).

Lemma 1.9. *Let L be a context-free language. Then there is an integer $p > 0$, depending only on L, such that, if $z \in L$ and $|z| \geq p$, then z can be written as $z = uvwxy$, where $|vwx| \leq p$, v and x are not both ε and for every $i \geq 0$, $uv^i wx^i y \in L$.*

Example. From an earlier example, $\{a^n b^n c^n \mid n > 0\}$ is type 1, but it is not of type 2 (context-free). For otherwise, Lemma 1.9 applies to $z = a^n b^n c^n$ for sufficiently large n, but no choices of v, x give $uv^i wx^i y \in L$ for all $i \geq 0$.

Thus there are strict inclusions of classes of languages:

$$\{\text{regular languages}\} \subsetneqq \{\text{context-free languages}\} \subsetneqq \{\text{context-sensitive languages}\}.$$

Eventually (see the note preceding Theorem 3.12), we shall show there is a type 0 language which is not type 1, so the inclusions in the Chomsky hierarchy of languages are all strict.

Although the class of regular languages is the most restricted class we have considered, regular languages are nevertheless important in computer science. We refer to [21, §2.8] and [22, §3.3] for a discussion of their uses, including lexical analysers and searching for strings. This involves another way of describing rational languages, by means of "rational expressions".

The ideas of rational expression, rational language and recognition by a FSA can be generalised, and there are notions of *rational* and *recognisable* subset of a monoid, leading to the idea of *star height* of a monoid. For a discussion of this, see [14] and [31]. The idea of an automaton over a subset A of a monoid is obtained by taking the labels on the transition diagram to be elements of A, so the automaton recognises a subset of the monoid generated by A. There is a generalisation of Theorem 1.6. If A is a set of generators for a monoid N, then a subset of N is rational if and only if it is recognised by an automaton over A. See [9, Theorem 2.6]. Also, the pushdown stack automata considered in Chap. 4 can be viewed as automata over a suitable monoid. See [9, §7].

We now turn to the class of machines which recognise type 0 languages.

Turing Machines. A Turing machine has a similar description to a FSA, but is allowed to do more. It can move the tape in both directions, print a new symbol on the scanned square, and extra symbols are allowed on the tape which are not part of the alphabet of the language recognised, which is called the input alphabet. Here is the formal definition.

Definition. A Turing machine is a sextuple $T = (Q, F, A, I, \tau, q_0)$, where

(1) Q is a finite set (the set of *states*);
(2) F is a subset of Q (the set of *final* states);
(3) A is a finite set (the *tape alphabet*) with a distinguished element B (the *blank symbol*);
(4) I is a subset of $A \setminus \{B\}$ (the *input alphabet*);
(5) $\tau \subseteq Q \times A \times Q \times A \times \{L, R\}$ (the set of *transitions*), where $\{L, R\}$ is a two-element set;
(6) $q_0 \in Q$ (the *initial state*).

Often, "Turing machine" will be abbreviated to "TM".

Thus the idea is that T has a read/write head scanning a tape divided into squares, each of which has a letter from A printed on it, and is in a certain state (element of Q). Elements of τ will be written without parentheses or commas. If $qaq'a'L \in \tau$, this means that, if T is in state q, reading a, it can change to state q', overwrite the scanned square with a' and move the head one square to the left (equivalently, move the tape one square to the right). If L is replaced by R, the head moves one square to the right.

No restriction is placed on the length of the tape, and it is convenient to view it as infinite in both directions, with all but finitely many squares blank (i.e. having B written on them).

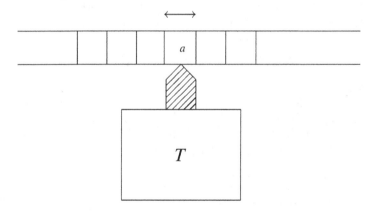

Figure 1.7

We now formalise this idea by giving a precise definition of a computation. Some preliminary definitions are required.

Definition. A *tape description* for a TM T as above is a triple (a, α, β) where $\alpha :$ $\mathbb{N} \to A$ and $\beta; \mathbb{N} \to A$ are functions with $\alpha(n) = B$ and $\beta(n) = B$ for all but finitely many $n \in \mathbb{N}$.

The idea is that a is on the square being scanned and the successive letters on the tape to the right of the scanned square are $\alpha(0), \alpha(1), \ldots$ Similarly, β records the letters to the left of the scanned square:

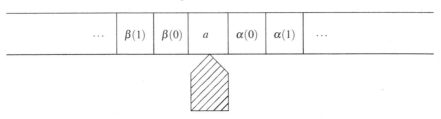

Figure 1.8

This is a good definition for theoretical purposes, but in practice it is useful to have a different way of giving a tape description. Suppose $\alpha(i) = B$ for $i > r$ and $\beta(i) = B$ for $i > l$. Then (a, α, β) is determined by the word

$$\beta(l)\beta(l-1)\ldots\beta(0)\underline{a}\alpha(0)\ldots\alpha(r)$$

with a underlined. Conversely, any word in A^+ with a letter underlined represents a tape description.

Definition. A *configuration* of T is a quadruple (q, a, α, β) where $q \in Q$ and (a, α, β) is a tape description.

Again, a configuration will sometimes be written as (q, w), where $q \in Q$ and w is a word in A^+ with a letter underlined.

We now describe the moves allowed by the transitions.

Definition. A configuration c' is obtained from a configuration c by a single move if one of the following holds:

(1) $c = (q, a, \alpha, \beta)$, $qaq'a'L \in \tau$ and $c' = (q', \beta(0), \alpha', \beta')$, where $\alpha'(0) = a'$, $\alpha'(n) = \alpha(n-1)$ for $n > 0$ and $\beta'(n) = \beta(n+1)$ for $n \geq 0$.
(2) $c = (q, a, \alpha, \beta)$, $qaq'a'R \in \tau$ and $c' = (q', \alpha(0), \alpha', \beta')$, where $\alpha'(n) = \alpha(n+1)$ for $n \geq 0$, $\beta'(0) = a'$ and $\beta'(n) = \beta(n-1)$ for $n > 0$.

It is now easy to define a computation.

Definition. A *computation* of T, starting at c and ending at c', is a finite sequence $c = c_1, \ldots, c_n = c'$ of configurations, where $n \geq 1$ and c_{i+1} is obtained from c_i by a single move, for $1 \leq i < n$.

We say that the computation *halts* if c' is a *terminal configuration*, that is, of the form (q, a, α, β), where no element of τ begins with qa.

Definition. $c \xrightarrow{T} c'$ means there is a computation of T, starting at c and ending at c'.

We can now define the language recognised by the TM. For $w = a_1 \ldots a_n \in A^*$, let $c_w = (q_0, \underline{a}_1 \ldots a_n) \; (= (q_0, \underline{B})$ if $w = \varepsilon)$.

Definition. The TM T *accepts* w if $c_w \xrightarrow{T} c'$ for some configuration $c' = (q, a, \alpha, \beta)$ such that $q \in F$.

The language *recognised* by T is

$$L(T) = \{w \in I^* \mid w \text{ is accepted by } T\}$$

(a language with alphabet I^* rather than A^*).

Deterministic Turing Machines. The requirements for a deterministic TM are less stringent than for a FSA. It is useful, even in a deterministic TM, to have the possibility of terminal configurations.

Definition. A TM T is *deterministic* if, for every pair $(q, a) \in Q \times A$, there is at most one element of τ which begins with qa.

For each configuration c of a deterministic TM, there is at most one configuration c' obtained from c by a single move. Put $\delta(c) = c'$, to obtain a partial function $\delta : C \to C$, where C is the set of configurations.

Note. A *partial function* $f : X \to Y$, where X and Y are sets, is a function $f : Z \to Y$, where Z is a subset of X. In this context, if $f : X \to Y$ is defined on all of X (i.e. $Z = X$), f is called a *total function*.

Further, if $qaq'a'd \in \tau$, we can write $q' = N_T(q, a)$, $a' = R_T(q, a)$ and $d = D_T(q, a)$, to obtain partial functions $N_T : Q \times A \to Q$, $R_T : Q \times A \to A$ and $D_T : Q \times A \to \{L, R\}$. (The subscript "$T$" will be needed in the next chapter when these functions are discussed simultaneously for a collection of TM's.) The next lemma is an immediate consequence of our definitions.

Lemma 1.10. *Let T be deterministic, $c = (q, a, \alpha, \beta)$ a configuration with $\delta(c)$ defined. Then*

(1) *If $D_T(q, a) = L$, then $\delta(c) = (N_T(q, a), \beta(0), \alpha', \beta')$, where $\alpha'(0) = R_T(q, a)$, $\alpha'(n) = \alpha(n-1)$ for $n > 0$ and $\beta'(n) = \beta(n+1)$ for $n \geq 0$.*
(2) *If $D_T(q, a) = R$, then $\delta(c) = (N_T(q, a), \alpha(0), \alpha', \beta')$, where $\alpha'(n) = \alpha(n+1)$ for $n \geq 0$, $\beta'(0) = R_T(q, a)$ and $\beta'(n) = \beta(n-1)$ for $n > 0$.*

$\square$

We can extend the definition of δ. Define $\bar{\delta} : C \times \mathbb{N} \to C$ by: $\bar{\delta}(c, 0) = c$, $\bar{\delta}(c, n+1) = \delta(\bar{\delta}(c, n))$. Note that $c \xrightarrow{T} c'$ if and only if $c' = \bar{\delta}(c, n)$ for some $n \geq 0$.

Given c, either $\bar{\delta}(c, n)$ is defined for all $n \geq 0$, or only for $0 \leq n \leq r$, where $r \geq 0$, meaning that the computation $c, \bar{\delta}(c, 1), \ldots, \bar{\delta}(c, r)$ halts. If $c = (q_0, a, \alpha, \beta)$

and this computation halts for some r, we say that T halts when started on the tape description (a, α, β).

As mentioned earlier, the class of languages recognised by a TM coincides with the class of type 0 languages. We shall prove only half of this now, deferring the converse until we give another characterisation of type 0 languages, in Chap. 3. (We shall also prove in Chap. 3 that a language recognised by a TM is actually recognised by a deterministic TM.) First, a remark is needed.

Remark 1.2. If $L = L(T)$ for some TM T, let T' be T with all transitions starting with qa, where $q \in F$ (the set of final states) removed. Then $L = L(T')$.

(For clearly $L(T') \subseteq L(T)$. If $c = c_1, \ldots, c_n$ is a computation of T, where c_n begins with a final state, taking n as small as possible subject to this gives a computation of T' starting at c. Hence $L(T) \subseteq L(T')$.)

Theorem 1.11. *If a language L is recognised by a TM, it is of type 0.*

Proof. Let L be recognised by $T = (Q, F, A, I, \tau, q_0)$. By Remark 1.2, we can assume that, if $q \in F$, no element of τ begins with qa.

Let G be the grammar (V_N, V_T, P, S), where $V_T = I$

$$V_N = ((I \cup \{\varepsilon\}) \times A) \cup Q \cup \{S, E_1, E_2, E_3\}$$

$(S, E_1, E_2, E_3$ being extra letters) and P consists of the following. (The reader is advised not to try to take in this list now, but to refer to it when needed in the commentary below.)

(1) $S \longrightarrow E_1 E_2$.
(2) $E_2 \longrightarrow (a, a)E_2$ for all $a \in I$.
(3) $E_2 \longrightarrow E_3$.
(4) $E_3 \longrightarrow (\varepsilon, B)E_3$, $E_1 \longrightarrow (\varepsilon, B)E_1$ (B is the blank symbol of T).
(5) $E_3 \longrightarrow \varepsilon$, $E_1 \longrightarrow q_0$.
(6) $q(a, C) \longrightarrow (a, D)p$, for all $qCpDR \in \tau$ and $a \in I \cup \{\varepsilon\}$.
(7) $(a, C)q \longrightarrow p(a, D)$, for all $qCpDL \in \tau$ and $a \in I \cup \{\varepsilon\}$.
(8) $(a, C)q \longrightarrow qaq$, $q(a, C) \longrightarrow qaq$ and $q \longrightarrow \varepsilon$, for all $a \in I \cup \{\varepsilon\}$, $C \in A$ and $q \in F$.

We show that $L = L_G$. Let $a_1 \ldots a_n \in I^*$; using productions (1), (2) and (3), we obtain

$$S \overset{\bullet}{\longrightarrow} E_1(a_1, a_1) \ldots (a_n, a_n)E_3$$

Suppose $a_1 \ldots a_n$ is accepted by T. In a computation starting at $c_0 = (q_0, \underline{a_1} \ldots a_n)$ and ending with a state in F, T uses only finitely many squares, say l, to the left of the initially scanned square. It also uses finitely many squares, say m, to the right of the square containing a_n, giving a block of $l + m + n$ squares. Now using (4) and (5),

$$S \overset{\bullet}{\longrightarrow} (\varepsilon, B)^l q_0(a_1, a_1) \ldots (a_n, a_n)(\varepsilon, B)^m$$

A configuration c in the computation can be described as $c = (q, x_1 \ldots \underline{x_i} \ldots x_{m+n+l})$, where $x_1, \ldots x_{m+n+l}$ are the letters currently on the initial block of squares. Associated to c is the word

$$\tilde{c} = (b_1, x_1) \ldots (b_{i-1}, x_{i-1}) q(b_i, x_i) \ldots (b_{m+n+l}, x_{m+n+l})$$

where $b_1 = \ldots = b_l = \varepsilon$, $b_{l+1} = a_1, \ldots b_{l+n} = a_n$, $b_{l+n+1} = \ldots = b_{l+n+m} = \varepsilon$. It follows by induction on the number of moves in the computation to get from c_0 to c that, using (6) and (7),

$$\tilde{c}_0 = (\varepsilon, B)^l q_0(a_1, a_1) \ldots (a_n, a_n)(\varepsilon, B)^m \overset{\bullet}{\longrightarrow} \tilde{c}$$

hence $S \overset{\bullet}{\longrightarrow} \tilde{c}$. If c is the last configuration in the computation, then $q \in F$, and by use of (8) we find $\tilde{c} \overset{\bullet}{\longrightarrow} a_1 \ldots a_n$, hence $S \overset{\bullet}{\longrightarrow} a_1 \ldots a_n$. Thus $L(T) \subseteq L_G$.

For the reverse inclusion, suppose $S \overset{\bullet}{\longrightarrow} a_1 \ldots a_n$. A corresponding derivation must start by deriving $(\varepsilon, B)^l q_0(b_1, b_1) \ldots (b_k, b_k)(\varepsilon, B)^m$ for some k, l, m, b_i, then continue using (6) and (7) (after possibly changing the places where the productions $E_3 \longrightarrow (\varepsilon, B)E_3$ and $E_3 \longrightarrow \varepsilon$ are used). Each use of (6) and (7) corresponds to a move of T, so we obtain a computation of T starting at $(q_0, \underline{b}_1 \ldots b_k)$. Eventually the word derived must contain some $q \in F$, in order to use (8), so T accepts $b_1 \ldots b_k$. The rest of the derivation can use only (8), resulting eventually in $b_1 \ldots b_k$. Thus $b_1 \ldots b_k = a_1 \ldots a_n$ is accepted by T. Therefore $L(T) = L_G$. $\qquad\square$

For a direct proof of the converse of Theorem 1.11, see [20, Theorem 7.3]. Before ending, we briefly describe the machines recognising context-sensitive languages. These will not be studied in detail. For a proof that these recognise exactly the context sensitive languages, see [20, §8.2].

A *linear bounded automaton* is a TM $T = (Q, F, A, I, \tau, q_0)$ such that only the part of the tape on which the input word is written may be used. More precisely

(1) The input alphabet I includes two special letters $\in$ and $\$$ (the left and right end markers of the tape);
(2) there are no transitions of the form $q \in q' aL$ or $q \$ q' aR$ (the read head cannot move beyond the end markers);
(3) the only transitions beginning $q \in$ (resp. $q \$$) have the form $q \in q' \in R$ (resp. $q \$ q' \$ L$) ($T$ cannot overprint $\in$ or $\$$).

A word $w \in (I \setminus \{\in, \$\})^*$ is accepted by T if $(q_0, \in w \$) \overset{}{\underset{T}{\longrightarrow}} (q, c)$ for some configuration c and $q \in F$. We define $L(T)$ to be the set of words accepted by w. It can be shown that a language L is context-sensitive if and only if it is $L(T)$ for some linear bounded automaton T. See [20], Theorems 8.1 and 8.2, or [21], Theorems 9.7 and 9.8. A language is called *deterministic context-sensitive* if it is $L(T)$ for some linear bounded automaton T which is deterministic as a TM. It is still unknown whether or not a context-sensitive language is deterministic context-sensitive.

Exercises on Chapter 1

1. Let G be the grammar $(\{S, A\}, \{a, b\}, P, S)$, where P consists of:

$$S \longrightarrow ab$$
$$S \longrightarrow aASb$$
$$A \longrightarrow bSb$$
$$AS \longrightarrow b$$

Is $ababbabb \in L_G$?

2. Let G be the grammar $(\{S\}, \{a, b, c\}, P, S)$, where P consists of $S \to aaS$ and $S \to bc$. Describe explicitly, with brief justification, the language L_G.

3. Draw the transition diagram for a FSA which recognises the language

$$\{(ab)^n \mid n = 0, 1, 2, \ldots\}.$$

4. Do the same for the language $\{vaw \mid v, w \in \{a, b\}^*\}$.

5. Let $L = \{1^m 01^n 01^{m+n} \mid m, n \in \mathbb{N}\}$, a language with alphabet $\{0, 1\}$.

   (i) Is L context-free (type 2)?
   (ii) Is L regular (type 3)?

6. Show that the language $\{a^i b^n c^n \mid n \geq 1, \ i \geq 0\}$ is context-free. Give an example of a pair of context-free languages L_1 and L_2 (on the same alphabet) such that $L_1 \cap L_2$ is not context-free.

Chapter 2
Recursive Functions

In this chapter, we consider the notion of a computable function $f : \mathbb{N}^n \to \mathbb{N}$. Such a function is computable if there is a finite set of instructions for a procedure which, if followed on input $(x_1, \ldots, x_n)$, terminates with output $f(x_1, \ldots, x_n)$ (for example, a computer program). No restriction is made on the time or space required in the device used to implement the procedure. (This, of course, is unrealistic, but it is easier to develop a theory without such restrictions.)

More generally, we consider *partial functions* $f : \mathbb{N}^n \to \mathbb{N}$. Recall that this means f is a function $X \to \mathbb{N}$ where X is a subset of $\mathbb{N}^n$. Such a function is computable if such a set of instructions exists, but the procedure terminates with output $f(x_1, \ldots, x_n)$ if this is defined, and otherwise does not terminate. (Also recall that, in this context, a function $f : \mathbb{N}^n \to \mathbb{N}$, defined on all of $\mathbb{N}^n$, is called a *total* function.)

This idea of computability cannot be subjected to a mathematical analysis because various terms, such as "procedure" have not been precisely defined. Nevertheless, it is possible to make some progress even with such a vague notion. As an example, take the following statement. Suppose $g, h : \mathbb{N} \to \mathbb{N}$ are two computable functions; then their composition $g \circ h$ is computable (recall that $(g \circ h)(n) = g(h(n))$). To "prove" this, take a procedure which computes h, give it input n, then pass the output to a procedure which computes g. The output is $g(h(n))$, so this is a procedure to compute $g \circ h$.

To obtain a mathematical theory, the way to proceed is to develop this idea. Write down a collection of functions which one expects to be computable (under any reasonable definition). Then give ways of constructing new functions which, applied to computable functions, should lead to new computable functions (such as composition, as described above). Then take all functions obtained from the initial functions by repeated use of these operations. This is what we shall do, leading to a class of functions called *partial recursive functions*.

We have to hope that we have written down enough initial functions and ways of constructing new functions that all possible computable functions are recursive. This is, of course, impossible to prove. Nevertheless, the assertion that this is true has a name.

I. Chiswell, *A Course in Formal Languages, Automata and Groups*,
DOI 10.1007/978-1-84800-940-0_2,
© Springer-Verlag London Limited 2009

Church's Thesis. The partial computable functions as described above are precisely the partial recursive functions.

This is sometimes called the Church-Turing thesis. In practice, it is used like the word "clearly" in other branches of mathematics. Thus "f is partial recursive by Church's thesis" means "f is obviously computable and I don't want to write out the lengthy details needed to prove it's recursive". We shall not use it in this way.

As evidence for Church's thesis, we shall consider several precise notions of computability and show that, in all cases, the partial computable functions are precisely the partial recursive functions. We shall consider computability by register programs. These resemble programs in a very simple assembly language. The machine which executes these programs has "registers", each of which can store any natural number, which can be changed when the program runs. This unrealistic assumption is compounded by making no limit on the number of registers a program may use, so the machine is given infinitely many registers. This reflects the statement made above in introducing computable functions: no restriction is made on the time or space required. Thus the machine implementing the program is expected to continue indefinitely without running out of power or breaking down.

We also consider computability by *abacus machines*, which can be viewed as versions of register programs written in a higher-level language, where only well-structured programs are possible. Finally, we discuss computability by Turing machines, a new use for them after their use in language recognition in Chap. 1.

Before defining the class of partial recursive functions, it is useful to define a smaller class, the class of *primitive recursive functions*, which are all total. Many standard functions on the natural numbers are primitive recursive. The definition involves just two ways of constructing new functions; a generalisation of composition discussed above, and "primitive recursion". We shall define these for arbitrary partial functions, so that no modification is needed when defining the partial recursive functions. Also, it is convenient to introduce the idea of a primitively recursively closed class, so that our results apply to other classes, such as the class of recursive functions defined later on.

Definition. Let $g : \mathbb{N}^r \to \mathbb{N}$, $h_1, \ldots, h_r : \mathbb{N}^n \to \mathbb{N}$ be partial functions. The function $f = g \circ (h_1, \ldots, h_r)$ obtained from $g, h_1, \ldots, h_r$ by *composition* is the partial function $f : \mathbb{N}^n \to \mathbb{N}$ defined by

$$f(x_1, \ldots, x_n) = g(h_1(x_1, \ldots, x_n), \ldots, h_r(x_1, \ldots, x_n))$$

where the left-hand side of the equation is defined if and only if the right-hand side is.

If $g, h_1, \ldots, h_r$ are computable functions, one can see that f is computable by a simple generalisation of the discussion above.

Definition. Let $g : \mathbb{N}^n \to \mathbb{N}$, $h : \mathbb{N}^{n+2} \to \mathbb{N}$ be partial functions. The function $f : \mathbb{N}^{n+1} \to \mathbb{N}$ obtained from g and h by *primitive recursion* is defined by:

$$f(x_1,\ldots,x_n,0) = g(x_1,\ldots,x_n)$$
$$f(x_1,\ldots,x_n,y+1) = h(x_1,\ldots,x_n,y,f(x_1,\ldots,x_n,y)).$$

For a formal proof that these equations do define a unique partial function f, we refer to [4, §3.7]. For given $(x_1,\ldots,x_n)$, $f(x_1,\ldots,x_n,y)$ is defined either for no y, for all y, or for $0 \le y \le r$ for some r. Note that $n = 0$ is allowed, when g is viewed as a fixed natural number.

If g and h are computable, then so is f. Given $\underline{x} = (x_1,\ldots,x_n)$, we first use a procedure to compute $g(\underline{x})$. If it terminates, the value obtained is $f(\underline{x},0)$. We can then use this value and a procedure to compute h to find $f(\underline{x},1)$. If this terminates, we can then use the computed value of $f(\underline{x},1)$ and the procedure to compute h to compute $f(\underline{x},2)$, and so on.

We also define the *initial functions* to be the functions in the following list:

(zero function) $z : \mathbb{N} \to \mathbb{N}$ defined by $z(x) = 0$ for all $x \in \mathbb{N}$
(successor function) $\sigma : \mathbb{N} \to \mathbb{N}$ defined by $\sigma(x) = x+1$
the *projection functions* $\pi_{in} : \mathbb{N}^n \to \mathbb{N}$ defined by $\pi_{in}(x_1,\ldots,x_n) = x_i$ (for $n \ge 1$ and $1 \le i \le n$).

The initial functions are all computable; it is left to the reader to justify this. We now define

$$\mathcal{P} = \{f \mid \text{for some } n > 0, f \text{ is a partial function } \mathbb{N}^n \to \mathbb{N}\}$$
$$\text{and} \quad \mathcal{T} = \{f \in \mathcal{P} \mid f \text{ is total}\}$$

In this chapter, a *class of functions* means a subset of $\mathcal{P}$ and a *class of total functions* means a subset of $\mathcal{T}$.

Definition. A class of total functions $\mathcal{C}$ is *primitively recursively closed* if

(1) $\mathcal{C}$ contains all the initial functions;
(2) $\mathcal{C}$ is closed under composition (i.e. if f is obtained from $g,h_1,\ldots,h_r$ by composition, and $g,h_1,\ldots,h_r$ are all in $\mathcal{C}$, then $f \in \mathcal{C}$);
(3) $\mathcal{C}$ is closed under primitive recursion (i.e. if f is obtained from g and h by primitive recursion, and $g, h \in \mathcal{C}$, then $f \in \mathcal{C}$).

There is a smallest primitively recursively closed class (the intersection of all primitively recursively closed total classes), called the class of *primitive recursive functions*.

Note. It is left to the reader to show that a function f is primitive recursive if and only if there is a sequence $f_0,\ldots,f_k = f$ of functions, where each f_i is either an initial function, or is obtained by composition from some of the f_j, for $j < i$, or is obtained by primitive recursion from two of the f_j with $j < i$. Such a sequence is called a *primitive recursive definition* of f.

Examples of Primitive Recursive Functions.

(1) (addition) The function $s : \mathbb{N}^2 \to \mathbb{N}$ defined by $s(x,y) = x+y$ is primitive recursive. For

$$s(x,0) = g(x) \text{ where } g = \pi_{11} \text{ (the identity mapping on } \mathbb{N})$$
$$s(x,y+1) = s(x,y) + 1 = h(x,y,s(x,y)), \text{ where } h = \sigma \circ \pi_{33}$$

so π_{11}, π_{33}, σ, $\sigma \circ \pi_{33}$, s is a primitive recursive definition.

(2) (multiplication) $m : \mathbb{N}^2 \to \mathbb{N}$ defined by $m(x,y) = xy$ is primitive recursive. For

$$m(x,0) = 0 = z(x)$$
$$m(x,y+1) = m(x,y) + x = s(\pi_{33}(x,y,m(x,y)), \pi_{13}(x,y,m(x,y)))$$
$$= h(x,y,m(x,y)), \text{ where } h = s \circ (\pi_{33}, \pi_{13}).$$

From this, it is easy to write down a primitive recursive definition. (In this and subsequent examples, this will be left to the reader.)

(3) (exponential function) $\exp(x,y) = x^y$ is primitive recursive. For

$$\exp(x,0) = 1$$
$$\exp(x,y+1) = m(x, \exp(x,y)).$$

(4) (factorial) $\mathrm{Fac}(x) = x!$ is primitive recursive since $\mathrm{Fac}(0) = 1$, $\mathrm{Fac}(x+1) = m(x+1, \mathrm{Fac}(x))$.

(5) Any constant function $\mathbb{N}^n \to \mathbb{N}$ is primitive recursive. For $n = 1$, the constant function 0 is z, the constant function 1 is $\sigma \circ z$, the constant function 2 is $\sigma \circ (\sigma \circ z)$, etc. For general n, the constant function c is $c' \circ \pi_{1n}$, where $c' : \mathbb{N} \to \mathbb{N}$ is the constant function with value c.

(6) (predecessor) We define $\mathrm{Pred}(x)$ to be $x-1$ if $x > 0$ and $\mathrm{Pred}(0)$ to be 0. This is primitive recursive since $\mathrm{Pred}(0) = 0$, $\mathrm{Pred}(x+1) = x$.

(7) (proper subtraction) $x \mathbin{\dot{-}} y = \max\{x-y, 0\}$ is primitive recursive: $x \mathbin{\dot{-}} 0 = x$, $x \mathbin{\dot{-}} (y+1) = \mathrm{Pred}(x \mathbin{\dot{-}} y)$.

(8) (modulus) $|x-y| = (x \mathbin{\dot{-}} y) + (y \mathbin{\dot{-}} x)$ is primitive recursive.

(9) (sign) $\mathrm{sg}(x) = \begin{cases} 0 & \text{if } x = 0 \\ 1 & \text{if } x > 0 \end{cases}$ is primitive recursive, because $\mathrm{sg}(0) = 0$ and $\mathrm{sg}(x+1) = 1$.

Remark 2.1. If $f : \mathbb{N}^n \to \mathbb{N}$ is in $\mathcal{C}$ (a primitively recursively closed class) and $g : \mathbb{N}^m \to \mathbb{N}$ is defined by $g(x_1, \ldots, x_m) = f(y_1, \ldots, y_n)$, where each y_i is either a constant or x_j for some fixed j, then $g \in \mathcal{C}$. (For $g = f \circ (h_1, \ldots, h_n)$, where h_i is either a constant function or some π_{jm}.)

Lemma 2.1. *Let $\mathcal{C}$ be a primitively recursively closed class, and let $g : \mathbb{N}^{n+1} \to \mathbb{N}$ be in $\mathcal{C}$. Then the following functions are in $\mathcal{C}$.*

(1) $f_1 : \mathbb{N}^{n+1} \to \mathbb{N}$, *where* $f_1(x_1, \ldots, x_n, y) = \sum\limits_{t=0}^{y} g(x_1, \ldots, x_n, t)$.

(2) $f_2 : \mathbb{N}^{n+1} \to \mathbb{N}$, where $f_2(x_1,\ldots,x_n,y) = \prod_{t=0}^{y} g(x_1,\ldots,x_n,t)$.

Proof. Both f_1 and f_2 are obtained by primitive recursion from functions in $\mathcal{C}$, since

(1) $f_1(\underline{x},0) = g(\underline{x},0)$, $f_1(\underline{x},y+1) = f_1(\underline{x},y) + g(\underline{x},y+1)$.
(2) $f_2(\underline{x},0) = g(\underline{x},0)$, $f_2(\underline{x},y+1) = f_1(\underline{x},y) \cdot g(\underline{x},y+1)$.

$\square$

Predicates. A predicate $P(x_1,\ldots,x_n)$ of n variables is a statement concerning these variables which is either true or false. In our case, the variables stand for elements of $\mathbb{N}$. Such a predicate is determined by the set $\{\underline{x} \in \mathbb{N}^n \mid P(\underline{x})$ is true$\}$ (and in formal approaches to set theory, would be identified with this set).

Recall that, if $A \subseteq \mathbb{N}^n$, the characteristic function of A is the function

$$\chi_A : \mathbb{N}^n \to \{0,1\} \text{ defined by } \chi_A(\underline{x}) = \begin{cases} 1 & \text{if } \underline{x} \in A \\ 0 & \text{if } \underline{x} \notin A. \end{cases}$$

If P is a predicate, χ_P is defined to be χ_A, where $A = \{\underline{x} \in \mathbb{N}^n \mid P(\underline{x})$ is true$\}$.

Definition. Let $\mathcal{C}$ be a primitively recursively closed class. A subset A of $\mathbb{N}^n$ is said to be *in* $\mathcal{C}$ if $\chi_A \in \mathcal{C}$. A predicate P of n variables is in $\mathcal{C}$ if $\{\underline{x} \in \mathbb{N}^n \mid P(\underline{x})$ is true$\}$ is in $\mathcal{C}$.

This is a somewhat awkward notation since "in" does not mean "is a member of". If $\mathcal{C}$ is the class of primitive recursive functions, we shall say A (or P) is primitive recursive, rather than A (or P) is in $\mathcal{C}$. Similar terminology will be used with the class of recursive functions defined later.

In the next lemma, the notation of propositional logic is used, and is assumed to be familiar. (Recall that $\wedge$ means "and", $\vee$ means "or" and $\neg$ means "not". Thus $P \vee Q$ is true, where P and Q are predicates, when either P is true, or Q is true, or both.)

Lemma 2.2. *Let $\mathcal{C}$ be a primitively recursively closed class. If A, $B \subseteq \mathbb{N}^n$ and A, B are in $\mathcal{C}$, then $A \cup B$, $A \cap B$ and $\mathbb{N}^n \setminus A$ are in $\mathcal{C}$. Consequently, if P, Q are predicates of n variables in $\mathcal{C}$, then $P \vee Q$, $P \wedge Q$ and $\neg P$ are in $\mathcal{C}$.*

Proof.

$$\chi_{A \cup B}(\underline{x}) = \chi_A(\underline{x}) \cdot \chi_B(\underline{x})$$
$$\chi_{A \cup B}(\underline{x}) = \text{sg}(\chi_A(\underline{x}) + \chi_B(\underline{x}))$$
$$\chi_{\mathbb{N}^n \setminus A}(\underline{x}) = 1 \dot{-} \chi_A(\underline{x})$$

$\square$

We next note that some familiar predicates of two variables are primitive recursive, for example $x = y$ (meaning the predicate $P(x,y)$ defined by $P(x,y)$ is true if and only if $x = y$).

Lemma 2.3. *The predicates $x = y$, $x \neq y$, $x \leq y$, $x < y$, $x \geq y$, $x > y$ are primitive recursive.*

Proof. Referring to the examples of primitive recursive functions given above, note that

$$\chi_{\neq}(x,y) = \mathrm{sg}(|x-y|), \; \chi_{<}(x,y) = \mathrm{sg}(x \,\dot{-}\, y)$$

and then use Lemma 2.2. In a slightly strange-looking notation, $=$ is $\neg(\neq)$, $\leq$ is $< \vee =$, $\geq$ is $\neg(<)$, etc. □

Bounded Quantifiers. These are quantifiers of the form $\exists y \leq z$ and $\forall y \leq z$, where y, z are variables representing elements of $\mathbb{N}$.

Lemma 2.4. *Let $\mathcal{C}$ be a primitively recursively closed class. If P is a predicate of $n+1$ variables in $\mathcal{C}$, then the predicates Q, R of $n+1$ variables defined below are in $\mathcal{C}$.*

(1) $Q(x_1,\ldots,x_n,z)$ *is true* $\Leftrightarrow \exists y \leq z(P(x_1,\ldots,x_n,y)$ *is true*);
(2) $R(x_1,\ldots,x_n,z)$ *is true* $\Leftrightarrow \forall y \leq z(P(x_1,\ldots,x_n,y)$ *is true*).

Proof. (1) $\chi_Q(\underline{x},z) = \mathrm{sg}\left(\sum_{y=0}^{z} \chi_P(\underline{x},y)\right)$;

(2) $\chi_R(\underline{x},z) = \prod_{y=0}^{z} \chi_P(\underline{x},y)$. Now use Lemma 2.1.

□

Bounded Minimisation. Let P be a predicate of $n+1$ variables. Define $f : \mathbb{N}^{n+1} \to \mathbb{N}$ by

$$f(\underline{x},z) = \begin{cases} \text{the least } y \leq z \text{ such that } P(\underline{x},y) \text{ is true} & \text{if such a } y \text{ exists} \\ z+1 & \text{otherwise.} \end{cases}$$

(Here $\underline{x} \in \mathbb{N}^n$.) The notation for this is $f(\underline{x},z) = \mu y \leq z P(\underline{x},y)$.

Lemma 2.5. *If $\mathcal{C}$ is a primitively recursively closed class and P is in $\mathcal{C}$, then f (as just defined) is in $\mathcal{C}$.*

Proof. This follows from Lemma 2.1, since

$$f(\underline{x},z) = \sum_{t=0}^{z}\prod_{y=0}^{t} \mathrm{sg}(1 \,\dot{-}\, \chi_P(\underline{x},y)).$$

□

Note. If $g : \mathbb{N}^{n+1} \to \mathbb{N}$ is is $\mathcal{C}$, then defining $P(\underline{x},z)$ to be true if and only if $g(\underline{x},z) = 0$, P is in $\mathcal{C}$ ($\chi_P(\underline{x},z) = 1 \,\dot{-}\, \mathrm{sg}(g(\underline{x},z))$). Thus, if $f(\underline{x},z) = \mu y \leq z(g(\underline{x},y) = 0)$, then f is in $\mathcal{C}$. On the other hand, every predicate P can be expressed in this way, with $g(\underline{x},z) = 1 \,\dot{-}\, \chi_P(\underline{x},z)$.

Definition by Cases. Let $f_1,\ldots,f_k : \mathbb{N}^n \to \mathbb{N}$ be in $\mathcal{C}$(a primitively recursively closed class) and let $P_1,\ldots,P_k$ be predicates in $\mathcal{C}$, of n variables. Suppose that for all $\underline{x} \in \mathbb{N}^n$, exactly one of $P_1(\underline{x}),\ldots,P_k(\underline{x})$ is true. Define $f : \mathbb{N}^n \to \mathbb{N}$ by

$$f(\underline{x}) = f_i(x) \quad \text{if } P_i(\underline{x}) \text{ is true, for } \underline{x} \in \mathbb{N}^n.$$

Lemma 2.6. *If f is so defined, then f is in $\mathcal{C}$.*

Proof. Just note that $f(\underline{x}) = f_1(\underline{x})\chi_{P_1}(\underline{x}) + \cdots + f_k(\underline{x})\chi_{P_k}(\underline{x})$. $\qquad\square$

Again, P_i can be given by: $P_i(\underline{x})$ is true if and only if $g_i(\underline{x}) = 0$, where g_i is in $\mathcal{C}$.

More Examples.

(1) The predicate of two variables, "x divides y" (written $x|y$) is primitive recursive. For $x|y \Leftrightarrow \exists t \le y(x.t = y)$. If $P(x,y,t)$ is the predicate $x.t = y$, then P is primitive recursive, as $\chi_P(x,y,t) = \chi_=(x.t,y)$.

(2) The predicate of one variable, "x is prime", is primitive recursive, for

$$x \text{ is prime} \Leftrightarrow (\neg \exists y \le x(1 < y \wedge y < x \wedge y|x)) \wedge (1 < x).$$

(3) The function $p(n) = $ the nth prime is primitive recursive. Since p has to be defined on $\mathbb{N}$, we let $p(0) = 2$, $p(1) = 3$, etc., so in fact $p(n)$ is the nth odd prime for $n > 0$. To prove p is primitive recursive, note that

$$p(n+1) = \text{least } p \text{ such that } (p(n) < p \text{ and } p \text{ is prime})$$

and this value of p is less than or equal to $p(n)! + 1$, since none of $p(0), \ldots, p(n)$ divide $p(n)! + 1$, but some prime does divide $p(n)! + 1$. Thus, if

$$f(x,y) = \mu p \le y(x < p \wedge (p \text{ is prime}))$$

then f is primitive recursive, and so is $h(x) = f(x, x! + 1)$. Since $p(0) = 2$, $p(n+1) = h(p(n))$, p is primitive recursive. In future, we prefer to write p_n rather than $p(n)$ for this function.

(4) Let $v(n,m)$ be the highest power of p_n dividing m. This does not make sense when $m = 0$, but we define v by

$$v(n,m) = \mu y \le m(\neg(p_n^{y+1} | m))$$

so v is primitive recursive. This gives $v(n,0) = 1$, which will not cause problems. If $p = p_n$, we define $\log_p : \mathbb{N} \to \mathbb{N}$ by $\log_p(m) = v(n,m)$, a primitive recursive function. Thus $\log_p$ is essentially the p-adic valuation (except that $\log_p(0) = 1$), rather than the logarithm function encountered in analysis.

(5) Define $\text{quo}(x,y) = \lfloor \frac{y}{x} \rfloor$ to be the quotient when y is divided by x. Then quo is primitive recursive. For $\text{quo}(x,0) = 0$, and

$$\text{quo}(x,y+1) = \begin{cases} \text{quo}(x,y) + 1 & \text{if } y + 1 = x(\text{quo}(x,y) + 1) \\ \text{quo}(x,y) & \text{otherwise.} \end{cases}$$

Thus, if we define

$$h(x,y,z) = \begin{cases} z+1 & \text{if } y+1 = x(z+1), \text{ i.e. } \underbrace{|(y+1)-x(z+1)| = 0}_{P(x,y,z)} \\ z & \text{otherwise, i.e. if } \neg P(x,y,z) \text{ is true} \end{cases}$$

then h is primitive recursive by Lemma 2.6 (with $P_1 = P$, $P_2 = \neg P$). Since $\mathrm{quo}(x,0) = 0$ and $\mathrm{quo}(x,y+1) = h(x,y,\mathrm{quo}(x,y))$, it follows that quo is primitive recursive. Note that if $x = 0, \lfloor \frac{y}{x} \rfloor$ is undefined, but this definition gives $\mathrm{quo}(0,y) = 0$ for all $y \in \mathbb{N}$.

(6) The remainder when y is divided by x

$$\mathrm{rem}(x,y) = y - x.\mathrm{quo}(x,y) = y \mathbin{\dot{-}} x.\mathrm{quo}(x,y)$$

is primitive recursive. Note that $\mathrm{rem}(x,y) = y$ if $x = 0$, otherwise $0 \leq \mathrm{rem}(x,y) < x$; also, $\mathrm{rem}(1,y) = 0$.

Iteration. Let X be a set, $f : X \to X$ a partial function. The *iterate* of f is the partial function $F : X \times \mathbb{N} \to X$ defined by $F(x,0) = x$, $F(x,n+1) = f(F(x,n))$ (so $F(x,n) = f^n(x)$ in the usual sense if f is total).

We have a notion of a function $f : \mathbb{N}^n \to \mathbb{N}$ being in a class $\mathcal{C}$. We can extend this to functions $f : \mathbb{N}^n \to \mathbb{N}^k$, by saying that f is in $\mathcal{C}$ if the coordinate functions $\pi_{ik} \circ f$ are in $\mathcal{C}$ for $1 \leq i \leq k$.

Definition. A class $\mathcal{C}$ of functions is *closed under iteration* if, whenever $f : \mathbb{N}^n \to \mathbb{N}^n$ is in $\mathcal{C}$, then its iterate $F : \mathbb{N}^{n+1} \to \mathbb{N}^n$ is in $\mathcal{C}$.

One can show that if $\mathcal{C}$ is primitively recursively closed, then $\mathcal{C}$ is closed under iteration (see Exercise 6 at the end of this chapter, or just refer to [4, p. 40]). However, we are interested in a kind of converse.

Lemma 2.7. *Let $\mathcal{C}$ be a class of functions which contains the initial functions and is closed under composition and iteration. Then $\mathcal{C}$ is closed under primitive recursion.*

Proof. Let $f : \mathbb{N}^{n+1} \to \mathbb{N}$ be obtained from g, h by primitive recursion, where g, h are in $\mathcal{C}$. For $\underline{x} \in \mathbb{N}^n$, let $\varphi(\underline{x},y,z) = (\underline{x}, y+1, h(\underline{x},y,z))$ and let Φ be the iterate of φ. Then $\Phi(\underline{x},0,g(\underline{x}),y) = (\underline{x}, y, f(\underline{x},y))$ (by induction on y). It is easy to see φ is in $\mathcal{C}$, so Φ is in $\mathcal{C}$. Since g is in $\mathcal{C}$ and $\mathcal{C}$ is closed under composition, it follows easily that f is in $\mathcal{C}$. $\square$

We now come to the more general classes of recursive and partial recursive functions. Their definition involves just one more way of constructing new functions, minimisation. Let $f : \mathbb{N}^{n+1} \to \mathbb{N}$ be a partial function. We can define a new function $g : \mathbb{N}^n \to \mathbb{N}$ by saying $g(\underline{x})$ is the least y such that $f(\underline{x},y) = 0$. However, since f is partial, this needs some clarification, and the definition is as follows.

Definition. The function obtained from f by minimisation is the partial function $g : \mathbb{N}^n \to \mathbb{N}$ defined by

$$g(\underline{x}) = \begin{cases} r & \text{if } f(\underline{x},r) = 0 \text{ and for } 0 \le s \le r, \ f(\underline{x},s) \text{ is defined and not } 0 \\ \text{undefined} & \text{otherwise.} \end{cases}$$

We write $g(\underline{x}) = \mu y(f(\underline{x},y) = 0)$. By contrast with the bounded minimisation considered earlier, g may be partial even if f is total.

Definition. The function g is obtained from f by *regular minimisation* if, additionally, f is total and for all $\underline{x} \in \mathbb{N}^n$, there exists y such that $f(\underline{x},y) = 0$. (The function g is then total.)

If f is computable in the informal sense described at the beginning of this section, then $g(\underline{x}) = \mu y(f(\underline{x},y) = 0)$ is computable. To compute $g(\underline{x})$, take a procedure to compute f and use it to successively compute $f(\underline{x},0), f(\underline{x},1), f(\underline{x},2), \ldots$ until a value r is reached with $f(\underline{x},r) = 0$, then output r. This procedure will continue indefinitely if either a value s is reached with $f(\underline{x},s)$ undefined (and none of $f(\underline{x},0), \ldots f(\underline{x},s-1)$ is zero), or if there is no value of r such that $f(\underline{x},r) = 0$. These are precisely the circumstances under which $g(\underline{x})$ is undefined.

A plausible way of defining g is to change the first clause as follows: $g(\underline{x}) = r$ if $f(\underline{x},r) = 0$ and for $0 \le s \le r$, $f(\underline{x},s)$ is either undefined or is defined and not equal to 0. However, the procedure just given will no longer work. If this clause applies and there is some $s < r$ with $f(\underline{x},s)$ undefined, the procedure will continue indefinitely without outputting r. In fact, there are examples where, using this definition, one can argue that g is not computable (see the end of §2.4, p.32 in [4]). This is why we have not used this as the definition.

We are now ready to define the idea of recursive function.

Definition. The class of recursive functions is the smallest class $\mathcal{C}$ of total functions which is primitively recursively closed and closed under regular minimisation. (That is, if f is in $\mathcal{C}$ and g is obtained from f by regular minimisation, then g is in $\mathcal{C}$.)

Note that there is such a smallest class, namely the intersection of all such classes $\mathcal{C}$. As indicated earlier, a subset A of $\mathbb{N}^n$ is called *recursive* if χ_A is recursive, and a predicate P of n variables is *recursive* if $\{\underline{x} \in \mathbb{N}^n \mid P(\underline{x}) \text{ is true}\}$ is recursive.

Thus the lemmas above concerning predicates in a primitively recursive class apply to recursive predicates.

The idea of a recursive subset of $\mathbb{N}^n$ is the formal version of a *decidable* set. This is a set A for which there is a finite set of instructions for a procedure which, given $\underline{x} \in \mathbb{N}^n$, decides in finitely many steps whether or not $\underline{x} \in A$. Even with such a vague idea, it should be clear that A is decidable if and only if χ_A is computable.

Definition. The class of partial recursive functions is the smallest class of partial functions which contains the initial functions and is closed under composition, primitive recursion and minimisation (in what should be an obvious sense).

The class of partial recursive functions which are total is primitively recursively closed and closed under regular minimisation, so contains the class of recursive functions. That is, a recursive function is partial recursive and total. The converse is true, but it is not obvious and will be proved later. Also, a primitive recursive function is recursive. We note some examples to show we have really extended the class of primitive recursive functions, and not all partial recursive functions are recursive.

Examples.

(1) Let $f(x) = \mu y(x(y+1) = 0) = \begin{cases} 0 & \text{if } x = 0 \\ \text{undefined} & \text{otherwise} \end{cases}$.

Clearly f is partial recursive but not total, so not recursive.

(2) For examples of recursive functions which are not primitive recursive, see [4, §3.6]. A particularly interesting example is the function $A : \mathbb{N}^2 \to \mathbb{N}$ now generally known as Ackermann's function. It is a simplified version of Ackermann's original function, and is defined by

$$A(0,y) = y+1$$
$$A(x+1,0) = A(x,1)$$
$$A(x+1,y+1) = A(x,A(x+1,y))$$

This is not a variant of primitive recursion, and A is not primitive recursive. But A is recursive, and it should be clear that A is computable, in the intuitive sense given at the beginning of the chapter. For proofs, see [5, §3.6.2].

Register Programs. Consider a machine having a number of registers, which are storage devices, each of which can store a non-negative integer. The machine can be given instructions to perform certain simple operations on registers. After finitely many steps, only finitely many registers are used, the others being clear (i.e. have 0 stored in them). However, there is no limit on the number that can be used. It is convenient to view the machine as having infinitely many registers, numbered $1, 2, 3, \ldots$, where only finitely many have a non-zero entry.

x_1	x_2	x_3	$\cdots$	x_n	0	0	$\cdots$
1	2	3		n			

Figure 2.1

The register contents are described by an infinite sequence $\underline{x} = (x_1, x_2, x_3, \ldots)$ of natural numbers, indexed by the positive integers, with $x_k = 0$ for all but finitely many values of k. Let Σ be the set of all such sequences.

Instructions are given to the machine by means of a *program*. We shall give a formal definition of a program, then indicate the intended meaning of the instructions.

Definition. A *register program P* is a finite sequence $\alpha_1, \ldots, \alpha_r$ where each α_i has the form $i.\beta$, and β_i is an *instruction*, that is, one of

$$a_k, \; s_k, \text{STOP}, \; J_k(l,m)$$

where $k \geq 1$ and $1 \leq l, m \leq r$. We also require that α_r is *terminal*, i.e. of the form $r.\text{STOP}$.

We call i the *label* on α_i, and α_i is called a *line* of the program. The intended meaning of the instructions is as follows:

   a_k: add 1 to the contents of register k;
   s_k: subtract one from the contents of register k, if this is not zero;
   $J_k(l,m)$: if register k is clear (i.e. contains 0), jump to instruction labelled l, otherwise to the instruction labelled m.

The instructions are executed in order unless a jump or STOP instruction is encountered. The STOP instruction means exactly what it says-when it is encountered no further instructions are carried out. Following the usual practice, the lines of a register program are written in a vertical list.

Example. Consider the program $O(k)$:

1. $J_k(4,2)$
2. s_k
3. $J_k(4,2)$
4. STOP

Starting at Line 1, if register k is clear we go to Instruction 4 and stop, otherwise go to Instruction 2 and subtract 1 from register k. Then we go to Instruction 3. If register k is now clear, we go to 4, otherwise back to 2. Thus, while register k is not clear, Instructions 2 and 3 are repeatedly executed until it is. That is, $O(k)$ clears the contents of register k.

   Given n, it is easy to construct a register program using more than n registers. Thus the machine which runs all register programs must have infinitely many registers. An alternative approach is to give each register program its own machine, with finitely many registers, sufficient to run the program. This would no longer be possible if we considered more complicated programs with instructions of the form "if r is the contents of register k, do something related to register r".

   We now give formal definitions of the effect that any register program P has on the registers of our machine.

Definition. A *configuration* of P is a pair $(i,\underline{x})$, where i is a label and $\underline{x} \in \Sigma$. It is *terminal* if the line labelled i is terminal, i.e. is $i.\text{STOP}$.

(The interpretation is that $\underline{x}$ represents the contents of the registers and the instruction of line i is about to be executed.) Given a non-terminal configuration $(i,\underline{x})$, carrying out Instruction i will result in a new configuration, which is described in the following definition.

Definition. If $(i,\underline{x})$ is a non-terminal configuration, the configuration $(j,\underline{y})$ *yielded* by $(i,\underline{x})$ is defined by:

(1) if line i has instruction a_k, then $j = i+1, y_p = \begin{cases} x_p & \text{if } p \neq k \\ x_p + 1 & \text{if } p = k \end{cases}$

(2) if line i has instruction s_k, then $j = i+1, y_p = \begin{cases} x_p & \text{if } p \neq k \\ x_p \dotminus 1 & \text{if } p = k \end{cases}$

(3) if line i has instruction $J_k(l,m)$, then $y = x$, $j = \begin{cases} l & \text{if } x_k = 0 \\ m & \text{otherwise} \end{cases}$

Definition. The *computation* of P starting from $\underline{x} \in \Sigma$ is the finite or infinite sequence

$$(i_1, \underline{x}_1), (i_2, \underline{x}_2), \ldots$$

where $i_1 = 1$, $\underline{x}_1 = \underline{x}$ and $(i_{q+1}, \underline{x}_{q+1})$ is the configuration yielded by $(i_q, \underline{x}_q)$, unless $(i_q, \underline{x}_q)$ is the last term in the sequence, in which case it must be terminal.

This defines a partial function $\varphi_P : \Sigma \to \Sigma$:

$$\underline{x}\,\varphi_P = \begin{cases} \underline{y} & \text{if the computation of } P \text{ starting from } \underline{x} \text{ is a finite sequence} \\ & \text{whose last term is } (i, \underline{y}) \text{ for some } i \\ \text{undefined} & \text{otherwise} \end{cases}$$

(It is convenient to write φ_P on the right, as will be seen later.) For example, if $P = O(k)$,

$$\underline{x}\,\varphi_P = (x_1, \ldots, x_{k-1}, 0, x_{k+1}, \ldots)$$

Using φ_P, P determines a partial function $\mathbb{N}^n \to \mathbb{N}$, for every $n \geq 1$.

Definition. The partial function $f : \mathbb{N}^n \to \mathbb{N}$ is *computed* by the register program P if

$$f(x_1, \ldots, x_n) = \begin{cases} y & \text{if } (x_1, \ldots, x_n, 0, 0, \ldots)\,\varphi_P = (y, \ldots) \\ \text{undefined} & \text{if } (x_1, \ldots, x_n, 0, 0, \ldots)\,\varphi_P \text{ is undefined} \end{cases}$$

(In the first case, f is given the value in register 1, and the values in the other registers are irrelevant.)

Abacus Machines. These are not really machines, but are words meant to represent certain (well-structured) register programs. The alphabet is $\{a_k, s_k, (,)_k \mid k \geq 1\}$, which is infinite, including an infinite collection of indexed right parentheses, $)_1,)_2$, etc. To each abacus machine is associated a natural number (its *depth*) and there is a notion of *simple* abacus machine. The definition is by induction on depth.

(1) a_k, s_k ($k \geq 1$) are the only simple abacus machines of depth 0.
(2) The abacus machines of depth n are the words $M_1 \ldots M_r$, where each M_i is a simple abacus machine of depth at most n, and some M_i has depth exactly n.
(3) The simple abacus machines of depth $n+1$ are the words $(M)_k$, where M is an abacus machine of depth n and $k \geq 1$.

An abacus machine is a set of instructions for operating on the registers, as follows.

a_k: add 1 to contents of register k; s_k: subtract 1 from register k unless it contains 0.

$M_1 \ldots M_r$: execute $M_1, \ldots, M_r$ in succession.

$(M)_k$ while $x_k \neq 0$ do M, where x_k is the contents of register k. (That is, check if $x_k \neq 0$ and if so, execute M. Do this repeatedly until $x_k = 0$.)

Note that the set of instructions corresponding to M may not terminate, for example $a_k(a_k)_k$ just keeps incrementing register k by 1. Here are some examples which carry out useful tasks.

Examples.

(1) $Clear_k = (s_k)_k$ (clears the contents of register k).
(2) $Descopy_{p,q} = Clear_q (s_p a_q)_p$ (copies contents of register p to register q and clears register p. This is short for "destructive copy" since the contents of register p are destroyed).
(3) $Copy_{p,q,r} = Clear_q (s_p a_q a_r)_p (s_r a_p)_r$ (if register r is clear, copies register p to register q, leaving registers other than q unchanged).

Next we prove some results on the structure of an abacus machine.

Lemma 2.8. (1) *An abacus machine has the same number of left and right parentheses (all the letters $)_k$, $k \geq 1$ are regarded as right parentheses here).*

(2) *A proper non-empty prefix of a simple abacus machine has more left than right parentheses (the proper non-empty prefixes of a word $u_1 \ldots u_m$ are $u_1 \ldots u_l$ where $1 \leq l < m$).*

Proof. We use induction on depth, when (1) becomes obvious. Clearly (2) holds for simple abacus machines of depth 0 (they have no proper non-empty prefixes). Suppose (2) holds for simple abacus machines of depth at most n, and let M be a simple abacus machine of depth $n + 1$. Then $M = (M_1 \ldots M_r)_k$, where each M_i is a simple abacus machine of depth at most n. The proper non-empty prefixes of M are $(M_1 \ldots M_{i-1} M_i'$, where M_i' is a prefix of M_i (possibly ε or M_i). By (1) and the induction hypothesis, $M_1, \ldots, M_{i-1}, M_i'$ all have at least as many left as right parentheses, hence so does $M_1 \ldots M_{i-1} M_i'$. Therefore $(M_1 \ldots M_{i-1} M_i'$ has more left than right parentheses.   □

Lemma 2.9. (1) *If a string S is an abacus machine, then there is exactly one value of r and one sequence of simple abacus machines $M_1, \ldots, M_r$ such that $S = M_1 \ldots M_r$.*

(2) *If S is a simple abacus machine, there is a unique k such that S is either a_k, s_k or $(M)_k$, where M is an abacus machine uniquely determined by S.*

Proof. (1) We can write $S = M_1 \ldots M_r$ for some simple abacus machines $M_1, \ldots, M_r$. By Lemma 2.8, M_1 is the shortest prefix of S (other than ε) having the same number of left and right parentheses. If $M_1 \neq S$, we can write $S = M_1 S'$ and similarly M_2 is the smallest prefix of S' having the same number of left and right parentheses. Continuing, this determines $M_1, \ldots, M_r$ (and r) uniquely.

(2) This is obvious since $)_k$ is the last letter of $(M)_k$, and $(M)_k = (M')_k$ implies $M = M'$ (deleting the first and last letters on each side).   □

An abacus machine M defines a partial function $\varphi_M : \Sigma \to \Sigma$, as follows.

(1) $\underline{x}\varphi_{a_k} = \underline{y}$, where $\quad y_i = \begin{cases} x_i & \text{if } i \neq k \\ x_i + 1 & \text{if } i = k \end{cases}$

(2) $\underline{x}\varphi_{s_k} = \underline{y}$, where $\quad y_i = \begin{cases} x_i & \text{if } i \neq k \\ x_i \dot- 1 & \text{if } i = k \end{cases}$

(3) If $M = M_1 \ldots M_r$ (M_i simple) then $\underline{x}\,\varphi_M = \underline{x}\,\varphi_{M_1} \cdots \varphi_{M_r}$.

(4) If $M = (M')_k$, then $\underline{x}\,\varphi_M = \underline{x}\,\varphi_{M'}^t$, where $t \in \mathbb{N}$ is chosen as small as possible such that $(\underline{x}\,\varphi_{M'}^t)_k$ (the kth entry of $\underline{x}\,\varphi_{M'}^t$) is zero ($\underline{x}\,\varphi_M$ is undefined if no such t exists).

This defines φ_M by induction on the depth of M, using Lemma 2.9. (Having defined φ_M for M of depth at most n, (4) then defines φ_M for simple abacus machines of depth $n+1$, then (3) defines it for all abacus machines of depth $n+1$.) Incidentally, (3) is the reason φ_M is written on the right, so that the order of the M_i does not have to be reversed, and we write φ_P (where P is a register program) on the right for consistency. As with register programs, an abacus machine determines a function $f : \mathbb{N}^n \to \mathbb{N}$ for each $n > 0$.

Definition. A partial function $f : \mathbb{N}^n \to \mathbb{N}$ is *computed* by the abacus machine M if: $f(x_1, \ldots, x_n)$ is defined if and only if $(x_1, \ldots, x_n, 0, 0, \ldots)\,\varphi_M$ is in which case

$$(x_1, \ldots, x_n, 0, 0, \ldots)\,\varphi_M = (f(x_1, \ldots, x_n), \ldots).$$

As with the definition of computable by a register program, the entries other than the first in $(f(x_1, \ldots, x_n), \ldots)$ are irrelevant, but we show that they can all be taken to be 0. For this, the idea of registers used by an abacus machine is needed.

Definition. The registers *used* by an abacus machine M are those whose numbers appear as subscripts in the machine. Thus

(1) a_k, s_k use only register k

(2) $M_1 \ldots M_r$ uses those registers used by M_i for at least one value of i

(3) $(M)_k$ uses register k and the registers used by M.

Since only finitely many subscripts appear in an abacus machine, an abacus machine uses only finitely many registers.

Remark. If $\underline{x}\varphi_M = \underline{y}$ and register i is not used by M, then $x_i = y_i$. (This is easily proved by induction on depth.)

Lemma 2.10. *If $f : \mathbb{N}^n \to \mathbb{N}$ is computed by an abacus machine, it is computed by an abacus machine M such that:*
$f(x_1, \ldots, x_n)$ is defined if and only if $(x_1, \ldots, x_n, 0, 0, \ldots)\,\varphi_M$ is, in which case

$$(x_1, \ldots, x_n, 0, 0, \ldots)\,\varphi_M = (f(x_1, \ldots, x_n), 0, 0, \ldots).$$

Proof. Let f be computed by M'. Choose m greater than or equal to the number of any register used by M', and with $m \geq n$. Then $M = M' Clear_2 \ldots Clear_m$ is the required machine. $\qquad\square$

The goal now is to show that register program computable, abacus machine computable and partial recursive are equivalent notions. The first step is the following, which implies that abacus machine computable functions are register program computable. The proof spells out the assertion that abacus machines are meant to represent certain register programs.

Lemma 2.11. *If M is an abacus machine, there is a register program P such that $\varphi_P = \varphi_M$ and the only* STOP *instruction of P is in the last line.*

Proof. (1) If $M = a_k$, take P to be $\begin{cases} 1.a_k \\ 2.\text{STOP} \end{cases}$ and if $M = s_k$, take P to be $\begin{cases} 1.s_k \\ 2.\text{STOP} \end{cases}$.

(2) Suppose $M = M_1 \ldots M_r$, where there exist register programs P_i such that $\varphi_{P_i} = \varphi_{M_i}$ $(1 \leq i \leq r)$ and P_i has only one STOP instruction. We show that there is a register program P with only one STOP instruction and $\varphi_P = \varphi_M$. For notational convenience, we treat only the case $r = 2$, leaving the modifications in the general case to the reader.

Let P_1 have labels $1, \ldots, n$ and P_2 have labels $1, \ldots, p$. Re-label the lines of P_2 as $n+1, \ldots, n+p$ and replace any jump instructions $J_k(l, m)$ by $J_k(n+l, n+m)$, to get a sequence of lines P_2'. Replace line n of P_1 by $n.J_1(n+1, n+1)$ (an unconditional jump to the line labelled $n+1$) to obtain P_1'. Let P be the concatenation $P_1' P_2'$; then P is a register program with only one STOP instruction, and $\varphi_P = \varphi_M$.

(3) Suppose $\varphi_{P'} = \varphi_{M'}$, where P' has r lines and one STOP instruction, and $k \geq 1$. We construct a register program P with one stop instruction and $\varphi_P = \varphi_{(M')_k}$. Increase all labels of P' by 1 and replace any jump instructions $J_q(l, m)$ by $J_q(l+1, m+1)$. Add a new first line, $1.J_k(r+2, 2)$, then remove the last line $(r+1.\text{STOP})$ and add two new lines: $\begin{cases} r+1.J_k(r+2, 2) \\ r+2.\text{STOP} \end{cases}$. This gives the required program P.

The lemma now follows by induction on the depth of the abacus machine M. $\qquad\square$

For example, if $M = Clear_k$, the program P with $\varphi_P = \varphi_M$ given by the proof is $O(k)$. We next show that partial recursive functions are abacus computable. Two technical lemmas about abacus computability are needed.

Remark. If $\underline{x}, \underline{y} \in \Sigma$ and $x_i = y_i$ for all i such that the abacus machine M uses register i, then $\underline{x}\varphi_M = \underline{y}\varphi_M$. This is easily proved by induction on the depth of M.

Lemma 2.12. *Let $f_1, \ldots, f_r : \mathbb{N}^n \to \mathbb{N}$ be abacus computable and let $p \geq 0$ be an integer. Then there is an abacus machine N such that, for all $\underline{x} \in \Sigma$,*

$$\underline{x}\varphi_N = (x_1, \ldots, x_n, x_{n+1}, \ldots, x_{n+p}, f_1(\underline{x}), \ldots, f_r(\underline{x}), \ldots)$$

where $f_i(\underline{x})$ means $f_i(x_1, \ldots, x_n)$.

Proof. Let the abacus machine M_i compute f_i $(1 \leq i \leq r)$. Choose an integer m greater than the number of any register used by any of the M_i. Put $M_i' =$

$Clear_{n+1}\ldots Clear_m M_i$. By the remark, for any $\underline{x} \in \Sigma$, $\underline{x}\varphi_{M_i'} = (f_i(x_1,\ldots,x_n),\ldots)$. Let M_i'' be the machine obtained from M_i' by increasing every subscript of M_i' by q, where $q = p+r+n$. Let

$$K_i = Copy_{1,q+1,q+n+1}\ldots Copy_{n,q+n,q+n+1}M_i''$$

Then for any $\underline{x} \in \Sigma$,

$$\underline{x}\varphi_{K_i} = (x_1,\ldots,\underbrace{x_n+p+r}_{q},f_i(x_1,\ldots,x_n),\ldots)$$

Now put $N_i = K_i Descopy_{q+1,n+p+i}$. Then $N = N_1 \ldots N_r$ is the required machine. $\quad\square$

Corollary 2.13. *Under the hyptheses of Lemma 2.12, there is an abacus machine M such that, for all $\underline{x} \in \Sigma$,*

$$\underline{x}\varphi_M = (f_1(x_1,\ldots,x_n),\ldots,f_r(x_1,\ldots,x_n),x_{n+1},\ldots,x_{n+p},\ldots).$$

Proof. Let N be as in Lemma 2.12 and put $q = p+r+n$. Then

$$M = N Descopy_{n+1,q+1}\ldots Descopy_{n+p,q+p}Descopy_{n+p+1,1}\ldots Descopy_{n+p+r,p,r+p}$$

is the required machine. $\quad\square$

Theorem 2.14. *Partial recursive functions are abacus computable.*

Proof. We show that the set of abacus computable functions contains the initial functions and is closed under composition, primitive recursion and minimisation. By definition, the class of partial recursive functions is then a subset, proving the theorem.

Now $Clear_1$ computes the zero function, a_1 the successor function, $Descopy_{k,1}$ ($k \neq 1$) computes π_{kn} and $a_1 s_1$ computes π_{1n}, so the initial functions are abacus computable.

Suppose $f_1,\ldots,f_r : \mathbb{N}^n \to \mathbb{N}$ and $g : \mathbb{N}^r \to \mathbb{N}$ are abacus computable. By Cor. 2.13, there is an abacus machine M such that

$$(x_1,\ldots,x_{n+1},\ldots)\varphi_M = (f_1(x_1,\ldots,x_n),\ldots,f_r(x_1,\ldots,x_n),x_{n+1},\ldots).$$

Let g be computed by the abacus machine M', and choose m greater than the number of any register used by M. Then

$$M Clear_{r+1}\ldots Clear_m M'$$

computes $g \circ (f_1,\ldots,f_r)$. Thus the set of abacus computable functions is closed under composition.

Let $f : \mathbb{N}^n \to \mathbb{N}^n$ be such that its coordinate functions $f_i = \pi_{in} \circ f$ are abacus computable for $1 \leq i \leq n$. We show that the iterate of f is abacus computable (meaning its coordinate functions are abacus computable). By Cor. 2.13, there is an abacus machine M such that

$$(x_1, \ldots, x_{n+1}, \ldots) \, \varphi_M = (f_1(x_1, \ldots, x_n), \ldots, f_n(x_1, \ldots, x_n), x_{n+1}, \ldots).$$

Let $M' = (Ms_{n+1})_{n+1}$. Then

$$(x_1, \ldots, x_{n+1}, \ldots) \, \varphi_{M'} = (f_1^k(x_1, \ldots, x_n), \ldots, f_n^k(x_1, \ldots, x_n), 0, \ldots).$$

provided the right-hand side is defined, where $k = x_{n+1}$. Using M' and $M' Descopy_{i,1}$ ($2 \leq i \leq n$), we see that the iterate of f is abacus computable. By Lemma 2.7, the class of abacus computable functions is closed under primitive recursion.

Finally, let $f : \mathbb{N}^{n+1} \to \mathbb{N}$ be abacus computable. By Lemma 2.12, there is an abacus machine M such that for all $\underline{x} \in \Sigma$,

$$(x_1, \ldots, x_{n+1}, \ldots) \, \varphi_M = (x_1, \ldots, x_{n+1}, f(x_1, \ldots, x_{n+1}), \ldots).$$

Let $M' = Clear_{n+1} M (a_{n+1} M)_{n+2} Descopy_{n+1,1}$. Then M' computes the function h given by $h(x_1, \ldots, x_n) = \mu y(f(x_1, \ldots, x_n, y) = 0)$. Thus the set of abacus computable functions is closed under minimisation, completing the proof. $\qquad\square$

The next result finishes the proof that abacus computable, register machine computable and partial recursive are equivalent.

Theorem 2.15. *If $f : \mathbb{N}^n \to \mathbb{N}$ is a partial function computed by a register program, then f is partial recursive.*

Proof. Let f be computed by the register program P with labels $1, \ldots, r$. The mapping $(i, \underline{x}) \mapsto 2^i \prod_{m \geq 1} p_m^{x_m}$ is a one-to-one mapping from the set of configurations of P into $\mathbb{N}$, and $2^i \prod_{m \geq 1} p_m^{x_m}$ is called the *code* of $(i, \underline{x})$. (Note that $g \in \mathbb{N}$ is a code if and only if $1 \leq \log_2 g \leq r$.) Define

$$In(x_1, \ldots, x_n) = 2 \prod_{1 \leq m \leq n} p_m^{x_m}$$

(the code of $(1, (x_1, \ldots, x_n, 0, 0, \ldots))$) and

$$Out(g) = \log_3(g)$$

(the contents of register 1 if g is a code). Also define $Next : \mathbb{N} \to \mathbb{N}$ by:

$$Next(g) = \begin{cases} g, \text{ if } g \text{ is not a code, or is the code of a terminal configuration} \\ \text{code of the configuration yielded by } (i, \underline{x}), \\ \text{where } g \text{ is the code of } (i, \underline{x}), \quad \text{otherwise} \end{cases}$$

In the second case, put $i = \log_2(g)$. Then

if line i of P is $i.a_k$, then $Next(g) = 2.g.p_k$

if line i of P is $i.s_k$, then $Next(g) = \begin{cases} 2.\text{quo}(p_k,g) & \text{if } \log_{p_k}(g) \neq 0 \\ 2.g & \text{if } \log_{p_k}(g) = 0 \end{cases}$

if line i of P is $i.J_k(l,m)$, then $Next(g) = \begin{cases} 2^m.\text{quo}(2^i,g) & \text{if } \log_{p_k}(g) \neq 0 \\ 2^l.\text{quo}(2^i,g) & \text{if } \log_{p_k}(g) = 0 \end{cases}$

Let $Comp$ be the iterate of $Next$. Finally, Let

$$Term(g) = \begin{cases} 1 & \text{if } g \text{ is the code of a terminal configuration} \\ 0 & \text{otherwise} \end{cases}$$

Then In, Out, $Next$, $Comp$ and $Term$ are primitive recursive (exercise).

Now, putting $\underline{x} = (x_1, \dots, x_n)$,

$$f(\underline{x}) = Out(Comp(In(\underline{x}),t))$$

for any t such that $Comp(In(\underline{x}),t)$ is the code of a terminal configuration, that is, such that $Term(Comp(In(\underline{x}),t)) = 1$ (and $f(\underline{x})$ is undefined if there is no such t). Put

$$F(\underline{x},t) = Out(Comp(In(\underline{x}),t))$$
$$G(\underline{x},t) = 1 \overset{\cdot}{-} Term(Comp(In(\underline{x}),t))$$

Then F and G are primitive recursive, and

$$f(\underline{x}) = F(\underline{x},t) \text{ for any } t \text{ such that } G(\underline{x},t) = 0$$
$$\text{(undefined if there is no such } t).$$

Hence

$$f(\underline{x}) = F(\underline{x}, \mu t(G(\underline{x},t) = 0)).$$

and it follows that f is partial recursive, being obtained from F and G by minimisation and composition. $\qquad\square$

Corollary 2.16. *For a partial function $f : \mathbb{N}^n \to \mathbb{N}$, the following are equivalent.*

(1) *f is partial recursive.*
(2) *f is abacus computable.*
(3) *f is computable by a register program.*

Proof. (1) $\Rightarrow$ (2) by Theorem 2.14, (2) $\Rightarrow$ (3) by Lemma 2.11, (3) $\Rightarrow$ (1) by Theorem 2.15. $\qquad\square$

We can now resolve a point from earlier in the chapter.

Corollary 2.17. *A partial function $f : \mathbb{N}^n \to \mathbb{N}$ is recursive if and only if it is partial recursive and total.*

Proof. It has already been noted that recursive implies partial recursive and total. If f is partial recursive, then it is computable by a register program (Cor. 2.16), and from the proof of Theorem 2.15, we can write $f(\underline{x}) = F(\underline{x}, \mu t(G(\underline{x}, t) = 0))$ for some primitive recursive functions F and G. If f is total, the minimisation must be regular, so f is recursive. $\square$

Computation of functions by Turing Machines. We show that the class of functions computable by a Turing machine coincides with the class of partial recursive functions. First, we have to specify how a TM computes a function, and this involves a special kind of TM.

Definition. A numerical TM is a deterministic TM $T = (Q, F, A, I, \tau, q_0)$ with $F = I = \emptyset$, $A = \{0, 1\}$ and $B = 0$ (blank symbol).

If $\underline{x} = (x_1, \ldots, x_n) \in \mathbb{N}^n$, define $Tape(\underline{x})$ to be the tape description $\underline{0}1^{x_1}01^{x_2}0\ldots01^{x_n}$. If T is a numerical TM, define $In_{T,n} : \mathbb{N}^n \to C$ (C is the set of configurations of T) by $In_{T,n}(\underline{x}) = (q_0, Tape(\underline{x}))$.

Definition. The partial function $\varphi_{T,n} : \mathbb{N}^n \to \mathbb{N}$ is defined by: if T, started on tape description $Tape(\underline{x})$ halts with the tape description $\underline{0}1^y = Tape(y)$ for some $y \in \mathbb{N}$ (i.e. the computation starting with $In_{T,n}(\underline{x})$, where $\underline{x} \in \mathbb{N}^n$, ends with a terminal configuration $(q, Tape(y)))$, then $\varphi_{T,n}(\underline{x}) = y$. Otherwise, $\varphi_{T,n}(\underline{x})$ is undefined.

The partial function $f : \mathbb{N}^n \to \mathbb{N}$ is called *TM computable* if $f = \varphi_{T,n}$ for some numerical TM T.

It is convenient to modify T. Add two new states p, h and the transitions

$$qapaL \quad \text{for all } (q, a) \in Q \times A \text{ such that no element of } \tau \text{ starts with } qa$$
$$\left.\begin{array}{l} pahaR \\ hapaL \end{array}\right\} \text{ for all } a \in A \text{ (i.e. } a = 0, 1).$$

Call the new machine T', let $Q' = Q \cup \{p, h\}$ be its set of states and let C' be its set of configurations. Then T' remains deterministic and transitions have the form

$$qaN_{T'}(q, a)R_{T'}(q, a)D_{T'}(q, a)$$

(see p. 17, just before Lemma 1.10) and $N_{T'}$, $R_{T'}$, $D_{T'}$ are defined on $Q' \times A$. Also, after suitable renaming, we can assume that

$$Q' = \{0, 1, \ldots, r-1\}, \; h = 0, \; p = 1, \; L = 0, \; R = 1.$$

Then $Q' \times A$ is a finite subset of $\mathbb{N}^2$, and putting $N_{T'}(x, y) = R_{T'}(x, y) = D_{T'}(x, y) = 0$ for $(x, y) \in \mathbb{N}^2 \setminus (Q' \times A)$, $N_{T'}$, $R_{T'}$, $D_{T'}$ are primitive recursive functions $\mathbb{N}^2 \to \mathbb{N}$.

Let $\delta : C' \to C'$ be the transition function of T', and let $\overline{\delta}$ be its iterate (these are total functions). If T has a computation ending with a terminal configuration $(q, Tape(y))$, then T', after two more moves, can enter state h without altering the tape. The only moves are then to alternate between states p and h, alternately moving the tape right and left. Note that $In_{T,n} = In_{T',n}$ and so

$$\varphi_{T,n}(\underline{x}) = \begin{cases} y, & \text{for any } t \text{ such that } \overline{\delta}(In_{T,n}(\underline{x}),t) = (h,\underline{0}1^y) \text{ for some } y \\ \text{undefined}, & \text{if there is no such } t \end{cases}$$

To show $\varphi_{T,n}$ is partial recursive, we need to code configurations by natural numbers. Define $Code : C' \to \mathbb{N}$ by $Code(q,a,\alpha,\beta) = 2^q 3^a 5^{\sigma(\alpha)} 7^{\sigma(\beta)}$ where

$$\sigma(\alpha) = \alpha(0) + \alpha(1)2 + \alpha(2)2^2 + \dots$$
$$\sigma(\beta) = \beta(0) + \beta(1)2 + \beta(2)2^2 + \dots$$

so $Code$ is a $1-1$ function.

Lemma 2.18. *There is a primitive recursive function* $Next : \mathbb{N} \to \mathbb{N}$ *such that*

$$Next(Code(c)) = Code(\delta(c))$$

for all $c \in C'$.

Proof. See Appendix A. $\square$

Now let $Comp$ be the iterate of $Next$, so

$$Comp(Code(c),t) = Code(\overline{\delta}(c,t))$$

which follows by an easy induction on t. Also,

$$Code(h,\underline{0}1^y) = 2^0 3^0 5^{1+2+2^2+\dots+2^{y-1}} 7^0 = 5^{2^y-1}$$

hence

$$\varphi_{T,n}(\underline{x}) = \log_2(1 + \log_5(Comp(Code(In_{T,n}(\underline{x})),t)))$$

for any t such that $Comp(Code(In_{T,n}(\underline{x}),t)) = Code(h,\underline{0}1^y)$ for some y (and is undefined if there is no such t). Define $\psi : \mathbb{N}^{n+1} \to \mathbb{N}$ by

$$\psi(\underline{x},t) = Comp(Code(In_{T,n}(\underline{x})),t))$$

so ψ is primitive recursive, since $\underline{x} \mapsto Code(In_{T,n}(\underline{x}))$ is primitive recursive (the proof is left to the reader). Further, the predicate P defined by

$$P(\underline{x},t) \text{ is true if and only if } \psi(\underline{x},t) = Code(h,\underline{0}1^y) = 5^{2^y-1} \text{ for some } y$$

is primitive recursive (again left as an exercise). Put

$$F(\underline{x},t) = \log_2(1 + \log_5(\psi(\underline{x},t)))$$
$$G(\underline{x},t) = 1 \dot{-} \chi_P(\underline{x},t)$$

so F and G are primitive recursive. Then $\varphi_{T,n}(\underline{x}) = F(\underline{x},t)$ for any t such that $G(\underline{x},t) = 0$ and is undefined if there is no such t. In particular,

$$\varphi_{T,n}(\underline{x}) = F(\underline{x},\mu t(G(\underline{x},t)=0))$$

is partial recursive. We have proved the following.

Theorem 2.19. *A TM computable function is partial recursive.*

□

We can take this further, by not only coding computations, but coding the TM's themselves. First, we need to order the transitions in a specific way. Given a linearly ordered set L, we can linearly order L^*. If $u = x_1 \ldots x_m$, $v = y_1 \ldots y_n \in L^*$, let $u < v$ if either $m < n$, or $m = n$ and there exists i such that $x_1 = y_1, \ldots, x_{i-1} = y_{i-1}$ but $x_i < y_i$. This is called the *ShortLex* ordering on L^*. Restricted to the set of words of a fixed length, it is called the *lexicographic* ordering.

Now let T be a numerical TM, and modify it as above to obtain T', so the transitions are words of length 5 in $\mathbb{N}^*$, and $\mathbb{N}$ is linearly ordered. We can therefore order the transitions by the lexicographic ordering, then number them to respect this ordering, say $q_i a_i q_i' a_i' D_i$ ($1 \le i \le k$) (so this is the ith transition in the lexicographic ordering, and k is the number of transitions). Define

$$gn(T') = 2^k 3^{q_0} \prod_{i=1}^{k} p_{5i}^{q_i} p_{5i+1}^{a_i} p_{5i+2}^{q_i'} p_{5i+3}^{a_i'} p_{5i+4}^{D_i}$$

(*gn* stands for "Gödel numbering", an idea discussed in the next chapter). Now define the following primitive recursive functions:

$$x(g, i) = \log_{p_{5i}}(g), \ y(g, i) = \log_{p_{5i+1}}(g)$$
$$k(g) = \log_2(g)$$
$$j = j(g, a, b) = \mu i \le k(g)(x(g, i) = a \wedge y(g, i) = b)$$
$$N(g, a, b) = \begin{cases} \log_{p_{5j+2}}(g) & \text{if } (\exists i \le k)(x(g, i) = a \wedge y(g, i) = b) \\ 0 & \text{otherwise} \end{cases}$$

Then if $g = gn(T')$, $N(g, a, b) = N_{T'}(a, b)$, where $N_{T'}$ is the function used in the proof of Lemma 2.18, defined just before Lemma 1.10. Similarly, we can define $R(g, a, b)$ and $D(g, a, b)$ (using $5j + 3$, $5j + 4$ instead of $5j + 2$). Also, we define

$$In_n(g, \underline{x}) = (\log_3(g), Tape(\underline{x})), \text{ for } g \in \mathbb{N}, \ \underline{x} \in \mathbb{N}^n.$$

In the proof of Lemma 2.18 (see Appendix A), replace $R_{T'}(a, b), N_{T'}(a, b), D_{T'}(a, b)$ by $R(g, a, b)$, etc., to get a primitive recursive function $NEXT : \mathbb{N}^2 \to \mathbb{N}$ such that, if $g = gn(T')$, $NEXT(g, x) = Next(x)$. Define $COMP : \mathbb{N}^3 \to \mathbb{N}$ by

$$COMP(g, x, 0) = x$$
$$COMP(g, x, t+1) = NEXT(g, COMP(g, x, t)).$$

Then *COMP* is primitive recursive, and if $g = gn(T')$, $COMP(g, x, t) = Comp(x, t)$, where *Comp* is the function in the proof of Theorem 2.19. This is proved by induction on t.

Put $\Psi(g,\underline{x},t) = COMP(g,Code(In_n(g,\underline{x})),t)$ and let $P(g,\underline{x},t)$ be the predicate

$$\Psi(g,\underline{x},t) = 5^{2^y-1} \text{ for some } y.$$

Let

$$F(g,\underline{x},t) = \log_2(1+\log_5(\Psi(g,\underline{x},t)))$$
$$G(g,\underline{x},t) = 1 \dot{-} \chi_P(g,\underline{x},t).$$

Then F and G are primitive recursive, and if $g = gn(T')$, then from the proof of Theorem 2.19

$$\varphi_{T,n}(\underline{x}) = \begin{cases} F(g,\underline{x},t) & \text{for any } t \text{ such that } G(g,\underline{x},t) = 0 \\ \text{undefined} & \text{if no such } t \text{ exists} \end{cases}$$

We have now proved the following.

Theorem 2.20. *For each $n \geq 1$, there are primitive recursive functions F, G :* $\mathbb{N}^{n+2} \to \mathbb{N}$ *such that, for any* TM *computable function $f : \mathbb{N}^n \to \mathbb{N}$, there exists $g \in \mathbb{N}$ such that, for all $\underline{x} \in \mathbb{N}^n$, $f(\underline{x}) = F(g,\underline{x},t)$ for any t such that $G(g,\underline{x},t) = 0$ and $f(\underline{x})$ is undefined if no such t exists. In particular,*

$$f(\underline{x}) = F(g,\underline{x},\mu t(G(g,\underline{x},t) = 0)).$$

$\square$

We shall prove that partial recursive functions are TM computable by showing that abacus computable implies TM computable. To do this, we need to construct some numerical TM's to perform specific tasks. There is quite a long list of them, but they provide insight into how TM's operate. A useful operation in their construction is the *product* of two TM's. This can only be defined when the first TM has a *halting state*.

Definition. Let T be any TM. A state h is called a *halting state* if, for any configuration $c = (q,a,\alpha,\beta)$, c is terminal if and only if $q = h$.

Products of TM's. Let T, T' be any TM's and assume T has a halting state. Rename the states so the halting state of T is also the initial state of T', and T, T' have no other states in common. Also, assume T, T' have the same blank symbol. Define TT' to be the TM whose states and transitions are those of T and T', and whose tape alphabet is the union of the tape alphabets of T and T'. The initial state and input alphabet are those of T, the final states are those of T'.

If $T_1,\ldots,T_r$ are TM's and $T_1,\ldots,T_{r-1}$ all have halting states, we define (recursively) $T_1 \ldots T_r = (T_1 \ldots T_{r-1})T_r$.

Some Numerical TM's

(1) P_0: this TM has set of states $Q = \{q_0,q,q'\}$ (where q_0 is the initial state) and four transitions

$$q_0aq0R, \quad qaq'aL \qquad (a = 0,1).$$

P_1: has the same set of states, but transitions $q_0 a q 1 R$, $q a q' a L$.

(For $i = 0$ or 1, P_i prints i on the scanned square and halts without moving the tape.)

(2) R: this has $Q = \{q_0, q\}$ and transitions $q_0 a q a R$ $(a = 0, 1)$.

L: has $Q = \{q_0, q\}$ and transitions $q_0 a q a L$ $(a = 0, 1)$.

(R, respectively L, moves one square right (resp. left) and halts.)

(3) R^*: $Q = \{q_0, q, q', h\}$, transitions

$$q_0 a q a R \ (a = 0, 1), \quad q 1 q 1 R, \quad q 0 q' 0 R, \quad q' a h a L \ (a = 0, 1).$$

(Moves to the first blank square to the right of the scanned square and halts.)

L^*: this is obtained from R^* by interchanging L and R in the transitions. (Moves to the first blank square to the left of the scanned square and halts.)

(4) *Test*: this has $Q = \{q_0, p_0, p_1, p_0', p_1'\}$ and transitions:

$$q_0 0 p_0' 0 R, \quad p_0' a p_0 a L \quad (a = 0, 1)$$
$$q_0 1 p_1' 1 R, \quad p_1' a p_1 a L \quad (a = 0, 1).$$

(*Test* leaves the tape description unaltered, changes to state p_0 (respectively p_1) if 0 (resp. 1) is scanned initially, and halts.)

(5) $Test\{T_0, T_1\}$: here T_0, T_1 are numerical TM's, with their states renamed so that they have no states in common, the initial state of T_i is p_i (the state of *Test* for $i = 0$, 1, and T_i, *Test* have only the state p_i in common. The states and transitions are those of T_0, T_1 and *Test*, and the initial state is that of *Test*.

(Started on a given tape description, this TM will follow the computation of T_0 or T_1, according to whether a 0 or 1 is initially scanned.)

(6) $Shiftleft = P_1 R^* L P_0 R$ (started on the tape description $u 0 0 1^x 0 v$, halts with the tape description $u 0 1^x 0 0 v$, for any $u, v \in \{0, 1\}^*$).

$Shiftright = P_1 L^* R P_0 L$ (started on $u 0 1^x 0 0 v$, halts with $u 0 0 1^x 0 v$).

(7) $Test_k = R^{* \, k-1} R Test\{L^{* \, k}, L^{* \, k}\}$. To describe the action of $Test_k$, define, for $\underline{x} \in \Sigma$

$$Tape(\underline{x}) = 0 1^{x_1} 0 1^{x_2} 0 1^{x_3} 0 \ldots$$

(*$Test_k$*, started on $Tape(\underline{x})$, halts with the same tape description, but in a state p_0 if $x_k = 0$, and in a state p_1 if $x_k \neq 0$.)

This completes our list, and we can now use these TM's to show that abacus computable implies TM computable.

Definition. A numerical TM T with a halting state *simulates* an abacus machine M if, for all $\underline{x} \in \Sigma$, T, when started on the tape description $Tape(\underline{x})$, halts if and only if $\underline{x} \varphi_M$ is defined, in which case it halts with the tape description $Tape(\underline{x} \varphi_M)$.

Theorem 2.21. *Any abacus machine M can be simulated by a numerical* TM *with a halting state.*

Proof. The proof is by induction on the depth of M. If M is a_k, M is simulated by $Add_k = Shiftleft^{k-1} P_1 L^{* \, k}$. If M is s_k, then M is simulated by

$$Sub_k = R^{* \, k-1} RTest\{L^{* \, k}, T_k\}$$

where $T_k = P_0 LShiftright^{k-1} R$.

If $M = M_1 \ldots M_r$ and T_i simulates M_i, then $T_1 \ldots T_r$ simulates $M_1 \ldots M_r$.

Suppose $M = (N)_k$ and T simulates N. Rename the states of T so its initial state is p_1 (a state of $Test_k$), its halting state is q_0 (the initial state of $Test_k$), but T and $Test_k$ have no other states in common. Let T' be the TM whose states and transitions are those of T and $Test_k$, with initial state q_0. Then T' simulates M. This is left to the reader (the halting state of T' is the state p_0 of $Test_k$). $\qquad \square$

Corollary 2.22. *If* $f : \mathbb{N}^n \to \mathbb{N}$ *is abacus computable, there exists a numerical* TM T *with a halting state such that, started on the tape description* $Tape(\underline{x})$ *(where* $\underline{x} \in \mathbb{N}^n$*),* T *halts if and only if* $f(\underline{x})$ *is defined, in which case* T *halts with the tape description* $\underline{0}1^y$*, where* $y = f(\underline{x})$.

A function is partial recursive $\Leftrightarrow$ *it is abacus computable* $\Leftrightarrow$ *it is* TM *computable.*

Proof. By Lemma 2.10, there is an abacus machine M such that $f(\underline{x})$ is defined if and only if $(x_1, \ldots, x_n, 0, 0, \ldots) \varphi_M$ is, for $x = (x_1, \ldots, x_n) \in \mathbb{N}^n$, in which case

$$(x_1, \ldots, x_n, 0, 0, \ldots) \varphi_M = (f(\underline{x}), 0, 0, \ldots).$$

By Theorem 2.21, there is a TM T which simulates M, and T is the required TM. Thus abacus computable implies TM computable, and the corollary follows by Cor. 2.16 and Theorem 2.19. $\qquad \square$

Our final result in this chapter makes use of this and Theorem 2.20.

Theorem 2.23 (Kleene Normal Form Theorem). *There exist primitive recursive functions* $\varphi : \mathbb{N} \to \mathbb{N}$ *and* $\psi : \mathbb{N}^3 \to \mathbb{N}$ *such that, if* $f : \mathbb{N} \to \mathbb{N}$ *is partial recursive, there exists* $g \in \mathbb{N}$ *such that*

$$f(x) = \varphi(\mu t(\psi(g, x, t) = 0)).$$

Proof. By Theorem 2.20 and Cor. 2.22, there are primitive recursive functions F, $G : \mathbb{N}^3 \to \mathbb{N}$ such that if $f : \mathbb{N} \to \mathbb{N}$ is partial recursive, there exists $g \in \mathbb{N}$ such that $f(x) = F(g, x, t)$ for any t such that $G(g, x, t) = 0$ (and $f(x)$ is undefined if no such t exists). Given f, choose such a number g.

Now put $\varphi = F \circ J_3^{-1}$ and

$$\psi(s, x, y) = GJ_3^{-1}(y) + |K(y) - s| + |KL(y) - x|$$

where J_3, K and L are as in Exercises (3) and (4) at the end of this chapter. Thus $J_3^{-1}(y) = (K(y), KL(y), LL(y))$, and φ, ψ are primitive recursive.

If $\psi(g, x, y) = 0$ then $K(y) = g$, $KL(y) = x$ and $G(g, x, t) = 0$, where $t = LL(y)$, so $f(x) = F(g, x, t) = \varphi(y)$.

Conversely, if $f(x)$ is defined, it is equal to $F(g,x,t)$ for some t with $G(g,x,t) = 0$. Put $y = J_3(g,x,t)$. Then $f(x) = \varphi(y)$ and $\psi(g,x,y) = 0$.

Thus $f(x)$ is defined if and only if there exists y with $\psi(g,x,y) = 0$, in which case $f(x) = \varphi(y)$ for any such y. In particular, $f(x) = \varphi(\mu t(\psi(g,x,t) = 0))$. $\square$

There are other ways of precisely defining computable functions, including several minor variants of register programs. The proof that computable functions are partial recursive often follows the method of proof used above for register program computable and TM computable. (Roughly, code computations by natural numbers, using primitive recursive functions.) The proof of the converse tends to use simulation (for example, of abacus machines by TM's).

For further reading on recursive function theory, see [29].

Exercises on Chapter 2

1. Show that the following functions are primitive recursive.

(a) $f(x_1,\ldots,x_n) = \max\{x_1,\ldots,x_n\}$.
(b) $f(x_1,\ldots,x_n) = \min\{x_1,\ldots,x_n\}$.
(c) $f(x) =$ the number of primes less than or equal to x.

2. Show that the binary predicate RP, where $\mathrm{RP}(x,y)$ means that x, y are relatively prime, is primitive recursive. Show that the function $\varphi : \mathbb{N} \to \mathbb{N}$, defined by $\varphi(x) =$ the number of positive integers less than or equal to x which are relatively prime to x, is primitive recursive.

3. We can define a bijection $J : \mathbb{N}^2 \to \mathbb{N}$ as follows. Write the elements of $\mathbb{N}^2$ as an infinite matrix:

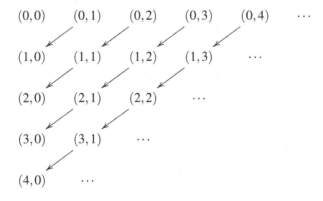

then write the entries as an infinite sequence by successively moving along the diagonals from northeast to southwest, as indicated by the arrows, giving

$$(0,0),(0,1),(1,0),(0,2),(1,1),(2,0),(0,3),(1,2),(2,1),(3,0),(0,4),(1,3),\ldots$$

Now there are $k+1$ pairs (m,n) with $m+n = k$, so the pair (m,n) occurs in the position $1+2+\ldots+(m+n)+m$ in the sequence (where the first position is numbered 0). We therefore define $J(m,n) = \frac{1}{2}(m+n)(m+n+1)+m$.

(a) Give a formal proof that J is bijective, and primitive recursive.
(b) Writing $J^{-1}(x) = (K(x),L(x))$, show that $K, L : \mathbb{N} \to \mathbb{N}$ are primitive recursive.

4. We can define bijections $J_n : \mathbb{N}^n \to \mathbb{N}$ for all n inductively by $J_1 = \pi_{11}, J_2 = J$ (the function in the previous exercises), $J_3(x_1,x_2,x_3) = J(x_1,J(x_2,x_3))$, and in general $J_{n+1}(x_1,\ldots,x_{n+1}) = J(x_1,J_n(x_2,\ldots,x_{n+1}))$. It follows easily by induction on n that J_n is primitive recursive for all n.

Show that $J_3^{-1}(r) = (K(r),KL(r),LL(r))$, and that for all n, J_n^{-1} is primitive recursive (meaning its coordinate functions are primitive recursive).

5. If $f : \mathbb{N}^n \to \mathbb{N}$ is a partial function with finite domain, show that f is partial recursive.

6. If $f : \mathbb{N}^n \to \mathbb{N}^n$ is in $\mathcal{C}$ (a primitive recursively closed class), prove that its iterate $F : \mathbb{N}^{n+1} \to \mathbb{N}^n$ is in $\mathcal{C}$. (Hint: this is easy for $n = 1$; in the general case, consider the iterate of $J_n \circ f \circ J_n^{-1}$.)

7. If $h : \mathbb{N} \to \mathbb{N}$ is primitive recursive, show that $\varphi : \mathbb{N}^3 \to \mathbb{N}$ defined by $\varphi(x,t,r) = h^{t \dotdiv r}(x)$ is primitive recursive.

8. Suppose $h, k : \mathbb{N} \to \mathbb{N}$ and $f : \mathbb{N}^2 \to \mathbb{N}$ are primitive recursive, and $g : \mathbb{N}^2 \to \mathbb{N}$ is defined by

$$g(x,0) = k(x)$$
$$g(x,t+1) = f(x,g(h(x),t)).$$

Prove that g is primitive recursive.

(Hint: without mentioning g, give a definition of a function $G(x,t,r)$ by primitive recursion, such that $G(x,t,r) = g(h^{t-r}(x),t)$ for $t \geq r$. Using this definition, show that for $t \geq r$, $G(x,t+1,r) = G(h(x),t,r)$. Then put $g(x,t) = G(x,t,t)$ and show g is given by the equations above.)

9. Construct a numerical TM T_1 which, started on the tape description $u01^a0\underline{0}1^c$, halts with the tape description $u01^a01^c\underline{0}$, for any $u \in \{0,1\}^*$ and natural numbers a, c.

10. Construct a numerical TM T_2 which, started on the tape description $u01^a01^b\underline{0}1^c$, halts with the tape description $u01^{a+1}01^{b-1}\underline{0}1^{c+1}$, provided $b > 0$, for any $u \in \{0,1\}^*$ and natural numbers a, b, c.

11. Construct a numerical TM T_3 which, started on the tape description $u01^a01^b\underline{0}$, halts with the tape description $u01^{a+b}01^b\underline{0}$, for any $u \in \{0,1\}^*$ and natural numbers a, b.

12. Construct a numerical TM T_4 which, started on the tape description $u01^a01^b\underline{0}$, halts with the tape description $u01^{a+b}\underline{0}$, for any $u \in \{0,1\}^*$ and natural numbers a, b.

13. Given a positive integer k, construct a numerical TM T_5 which, started on the tape description $u01^a01^b\underline{0}$, halts with the tape description $u01^{a+kb}\underline{0}$, for any $u \in \{0,1\}^*$ and natural numbers a, b.

14. Given a positive integer k, construct a numerical TM T_6 which, started on the tape description $01^{x_1}01^{x_2}\ldots01^{x_n}\underline{0}$, halts with the tape description

$$\underline{0}\,1^{x_1+x_2k+x_3k^2+\ldots+x_nk^{n-1}}$$

for any $n, x_1,\ldots,x_n > 0$, and which, when started on a blank tape, halts on a blank tape (i.e. it works when $n = 0$ as well). (Your machine should have a halting state.)

(Hint: in **9–14**, use machines already constructed and products of TM's. In some cases, you may need to identify the initial state of one machine with a state of another machine.)

Chapter 3
Recursively Enumerable Sets and Languages

We begin by discussing recursively enumerable subsets of $\mathbb{N}$. This formalises the idea of a *listable* set. This means a set A which is $f(\mathbb{N})$ for some computable function f, so the elements of A can be listed as $f(0), f(1), f(2), \ldots$ by some procedure with a finite set of instructions. We also consider the empty set to be listable.

Definition. A subset A of $\mathbb{N}$ is *recursively enumerable* (abbreviated to r.e.) if $A = f(\mathbb{N})$ for some recursive function $f : \mathbb{N} \to \mathbb{N}$, or $A = \emptyset$.

There are several other equivalent ways of saying a set is r.e. The partial characteristic function χ_{pA} of A is defined by:

$$\chi_{pA} = \begin{cases} 1 & \text{if } x \in A \\ \text{undefined} & \text{if } x \notin A. \end{cases}$$

Lemma 3.1. *For $A \subseteq \mathbb{N}$, the following are equivalent.*

(1) *A is recursively enumerable.*
(2) *A is the domain of some partial recursive function $g : \mathbb{N} \to \mathbb{N}$.*
(3) *the partial characteristic function χ_{pA} is partial recursive.*
(4) *there is a partial recursive function $f : \mathbb{N} \to \mathbb{N}$ such that $A = f(\mathbb{N})$.*
(5) *either $A = \emptyset$, or there is a primitive recursive function $f : \mathbb{N} \to \mathbb{N}$ such that $A = f(\mathbb{N})$.*

Proof. $(1) \Rightarrow (2)$. If $A = \emptyset$, $A = \text{dom}(g)$, where g is a partial recursive function with empty domain, for example $g(x) = \mu y(x + y + 1 = 0)$. If $A = f(\mathbb{N})$, where f is recursive, let $g(x) = \mu y(f(y) = x)$. Then $A = \text{dom}(g)$.

$(2) \Rightarrow (3)$. Assume (2). Then $\chi_{pA} = 1 \overset{.}{-} z.g$, where z is the zero function, so χ_{pA} is partial recursive, as it is obtained from g and primitive recursive functions by composition.

$(3) \Rightarrow (4)$. Let $f = \pi_{11} + (1 \overset{.}{-} \chi_{pA})$. (Recall that π_{11} is the identity function on $\mathbb{N}$.)

I. Chiswell, *A Course in Formal Languages, Automata and Groups*,
DOI 10.1007/978-1-84800-940-0_3,
© Springer-Verlag London Limited 2009

$(4) \Rightarrow (5)$. Assume (4). By Theorem 2.23, there are primitive recursive functions $u : \mathbb{N} \to \mathbb{N}$, $v : \mathbb{N}^2 \to \mathbb{N}$ such that $f(x) = u(\mu t(v(x,t) = 0))$. (We obtain v from the function ψ in 2.23 by fixing a suitable value of g.) If $A = \emptyset$ then (5) is true; otherwise choose $a_0 \in A$ and define $F : \mathbb{N}^2 \to \mathbb{N}$ by

$$F(x,n) = \begin{cases} u(\mu t \le n(v(x,t) = 0)) & \text{if } \exists t \le n(v(x,t) = 0) \\ a_0 & \text{otherwise.} \end{cases}$$

Then F is primitive recursive (this is left as an exercise) and $F(\mathbb{N}^2) = A$. Let $J : \mathbb{N}^2 \to \mathbb{N}$ be the primitive recursive bijection in Exercise 3 of Chapter 2. Then by Exercise 3, $F \circ J^{-1} = F \circ (K,L) : \mathbb{N} \to \mathbb{N}$ is primitive recursive, with image A.

$(5) \Rightarrow (1)$. This is obvious. $\square$

Recall that a subset A of $\mathbb{N}$ is recursive if χ_A is recursive, and this formalises the idea of a decidable set. If A is listable, it does not follow (at least, not obviously) that A is decidable. There is a procedure to list the elements of A, so if $n \in A$, it will appear eventually in the list. But we have no idea, in general, when it will appear, so if $n \notin A$, this procedure will not tell us that $n \notin A$. (In fact, we shall shortly exhibit a r.e. non-recursive set.) However, if $\mathbb{N} \setminus A$ is also listable, then A is decidable. Given n, just list the elements of A and the elements of $\mathbb{N} \setminus A$ and see which list n eventually appears in. (We are ignoring the extreme cases $A = \emptyset$, $A = \mathbb{N}$.) Conversely, if A is decidable, we can list both A and $\mathbb{N} \setminus A$ by computing χ_A. Here is the formal version of this.

Lemma 3.2. *A subset A of $\mathbb{N}$ is recursive if and only if both A and $\mathbb{N} \setminus A$ are r.e.*

Proof. Suppose A is recursive. Then $\chi_{pA}(x) = 1 + \mu y(\chi_A(x) = 1)$ and $\chi_{\mathbb{N} \setminus pA}(x) = 1 + \mu y(\chi_A(x) = 0)$, so these partial characteristic functions are partial recursive, hence A, $\mathbb{N} \setminus A$ are r.e. by Lemma 3.1.

Suppose A, $\mathbb{N} \setminus A$ are r.e. If $A = \mathbb{N}$ or $A = \emptyset$, χ_A is constant, so (primitive) recursive, hence we can assume $A = f(\mathbb{N})$, $\mathbb{N} \setminus A = g(\mathbb{N})$ with f, g recursive.

Define $\begin{cases} h(2x) = f(x) \\ h(2x+1) = g(x) \end{cases}$. Then h is recursive, since

$$h(x) = \begin{cases} f(\text{quo}(2,x)) & \text{if quo}(2,x) = 0 \\ g(\text{quo}(2,x)) & \text{if quo}(2,x) = 1 \end{cases}$$

and $h(\mathbb{N}) = f(\mathbb{N}) \cup g(\mathbb{N}) = \mathbb{N}$.

Define $\varphi(x) = \mu y(h(y) = x)$. The minimisation is regular, so φ is recursive. If $h(y) = x$, then $\begin{cases} x \in A & \text{if } y \text{ is even} \\ x \in \mathbb{N} \setminus A & \text{if } y \text{ is odd} \end{cases}$

Hence $x \in A$ if and only if $\varphi(x)$ is even, so $\chi_A(x) = 1 \overset{.}{-} \text{rem}(2, \varphi(x))$ is recursive.

$\square$

Lemma 3.3. *(1) If A, B are r.e. then so are $A \cup B$ and $A \cap B$.*
(2) If $F : \mathbb{N} \to \mathbb{N}$ is partial recursive and A is r.e., then $F(A)$ and $F^{-1}(A)$ are r.e. If F is total and A is recursive, then $F^{-1}(A)$ is recursive.

Proof. (1) Clearly $A \cup B$ is r.e. if A or B is empty. Otherwise, take recursive functions f, g such that $A = f(\mathbb{N})$, $B = g(\mathbb{N})$. Then $A \cup B = h(\mathbb{N})$, where h is defined as in the proof of Lemma 3.2, so h is recursive.

By Lemma 3.1, there are partial recursive functions $\varphi, \psi : \mathbb{N} \to \mathbb{N}$ such that $A = \mathrm{dom}(\varphi)$, $B = \mathrm{dom}(\psi)$. Then $\varphi + \psi$ is partial recursive with domain $A \cap B$, so $A \cap B$ is r.e. by Lemma 3.1.

(2) This is clear if $A = \emptyset$. Otherwise, we can write $A = \mathrm{dom}(\varphi)$, where $\varphi : \mathbb{N} \to \mathbb{N}$ is partial recursive, and $A = f(\mathbb{N})$, where f is recursive. Then $F^{-1}(A) = \mathrm{dom}(\varphi \circ F)$, $F(A) = (F \circ f)(\mathbb{N})$ are r.e. Hence, if F is total, $\mathbb{N} \setminus F^{-1}(A) = F^{-1}(\mathbb{N} \setminus A)$ is r.e. provided $\mathbb{N} \setminus A$ is, and (2) follows by Lemma 3.2.   □

Lemma 3.4. *Let $A \subseteq \mathbb{N}$. Then A is recursive and infinite if and only if there is a recursive function $f : \mathbb{N} \to \mathbb{N}$ such that $f(\mathbb{N}) = A$ and f is strictly increasing.*

Proof. Suppose $f(\mathbb{N}) = A$ where f is recursive and strictly increasing. Then by an easy induction on n, $f(n) \geq n$ for all $n \in \mathbb{N}$, so A is infinite. Further, $a \in A$ if and only if $\exists n \leq a(f(n) = a)$. This is a recursive predicate by Lemma 2.4, so χ_A is recursive.

Conversely, suppose A is recursive and infinite. Define $\varphi : \mathbb{N} \to \mathbb{N}$ by

$$\varphi(x) = \mu y(y > x \wedge \chi_A(y) = 1)$$

a partial recursive function since χ_A is recursive, and total since A is infinite. Hence φ is recursive by Cor. 2.17. Let Φ be the iterate of φ, and put $f(n) = \Phi(a_0, n)$, where a_0 is the least element of A. Then f is recursive since Φ is, and $f(\mathbb{N}) \subseteq A$ since $\varphi(\mathbb{N}) \subseteq A$. Also, f is strictly increasing.

It remains to show that if $a \in A$, then $a \in f(\mathbb{N})$. Since $a_0 = f(0)$, we can assume $a > a_0$. Since f is strictly increasing, there is n such that $f(n) < a \leq f(n+1)$. By definition of f, $f(n+1)$ is the least element of A greater than $f(n)$. Hence $a = f(n+1) \in f(\mathbb{N})$.   □

Recall from Lemma 2.23 that there exist primitive recursive functions $F : \mathbb{N} \to \mathbb{N}$, $G : \mathbb{N}^3 \to \mathbb{N}$ such that for any partial recursive function $f : \mathbb{N} \to \mathbb{N}$, there exists k such that $f(x) = F(\mu t(G(k, x, t) = 0))$ for $x \in \mathbb{N}$.

Put $U(k, x) = F(\mu t(G(k, x, t) = 0))$ and let $f_k(x) = U(k, x)$. Then $\{f_k \mid k \in \mathbb{N}\}$ is the set of all partial recursive functions of one variable.

Proposition 3.5. *The set of recursive functions of one variable is not r.e., that is*

$$\{k \mid f_k \text{ is recursive}\}$$

is not r.e.

Proof. Suppose it is r.e., so equal to $g(\mathbb{N})$ for some recursive function $g : \mathbb{N} \to \mathbb{N}$. Put $h(x) = f_{g(x)}(x) + 1 = U(g(x), x) + 1$. Then h is partial recursive ($h = \sigma \circ U \circ (g \circ \pi_{11}, \pi_{11})$) and total, so recursive by Cor. 2.17. Hence $h = f_{g(m)}$ for some m. But then $h(m) = f_{g(m)}(m) = f_{g(m)}(m) + 1$, a contradiction.   □

Proposition 3.6. *The set $A = \{x \in \mathbb{N} \mid U(x, x) \text{ is defined}\}$ is r.e. but not recursive.*

Proof. First, $A = \text{dom}(U \circ (\pi_{11}, \pi_{11}))$, so A is r.e. If A were recursive, then $\mathbb{N} \setminus A$ would be r.e., so $\chi = \chi_{p(\mathbb{N} \setminus A)}$ would be partial recursive, hence $\chi = f_m$ for some m. Then $m \notin A \Longleftrightarrow \chi(m)$ is defined $\Longleftrightarrow f_m(m)$ is defined $\Longleftrightarrow U(m,m)$ is defined $\Longleftrightarrow$ $m \in A$, a contradiction. $\square$

Proposition 3.7. *The set* $B = \{(k,x) \mid U(k,x) \text{ is defined}\}$ *is not recursive.*

Proof. If it were, the set A in Prop. 3.6 would be recursive, since $\chi_A(x) = \chi_B(x,x)$.
                                                                                              $\square$

Note. Props. 3.6 and 3.7 are related to the "halting problem" for Turing machines. This is discussed in Appendix B. Also, the arguments in Props. 3.5 and 3.6 are related to "Cantor's diagonal argument", which is discussed in Appendix C.

Gödel Numbering

We want to introduce the ideas of recursive language and recursively enumerable language, corresponding to informal ideas of decidable and listable language. The way to proceed is, given an alphabet A, to code A^* by natural numbers. (This is similar to what happened in Chapter 2, where computations of register programs and of TM's, as well as TM's themselves, were coded by natural numbers (Theorem 2.15, Theorem 2.19 and Theorem 2.20).) Then we use the notions of recursive and r.e. already defined for subsets of $\mathbb{N}$.

We shall show that the r.e. languages are precisely the type 0 languages defined in Chapter 1, and complete the proof that these coincide with the languages recognised by a TM.

Let X be a countably infinite set, $f : X \to \mathbb{N}$ a bijection.

Definition. A subset A of X is *recursive* (resp. *r.e*) relative to f if $f(A)$ is recursive (resp. r.e.).

Definition. A Gödel numbering of X is an injective mapping $\varphi : X \to \mathbb{N}$ such that $\varphi(X)$ is recursive.

One can similarly define recursive and r.e. subsets relative to a Gödel numbering φ. We have not done so explicitly because we shall see that these ideas are equivalent to recursive and r.e. relative to a suitable bijection f. These notions depend on the choice of φ, but in many cases various natural choices for φ lead to the same collections of recursive and r.e. subsets of X.

Given a Gödel numbering $\varphi : X \longrightarrow \mathbb{N}$, there is a strictly increasing recursive function g from $\mathbb{N}$ onto $\varphi(\mathbb{N})$, by Lemma 3.4. Then $f = g^{-1} \circ \varphi : X \to \mathbb{N}$ is bijective, so we can consider recursive and r.e. sets of X relative to f.

Lemma 3.8. *In these circumstances, let A be a subset of X. Then*

$$A \text{ is r.e. relative to } f \Longleftrightarrow \varphi(A) \text{ is r.e.}$$

$$\text{and} \quad A \text{ is recursive relative to } f \Longleftrightarrow \varphi(A) \text{ is recursive}$$

Further, A is recursive if and only if A and $X \setminus A$ are r.e.

Proof. The set A is r.e. relative to f if and only if $g^{-1}(\varphi(A))$ is r.e. By Lemma 3.3, if $\varphi(A)$ is r.e. then $g^{-1}(\varphi(A))$ is r.e., and if $g^{-1}(\varphi(A))$ is r.e. then $g(g^{-1}(\varphi(A)))$ is r.e. Since g maps onto $\varphi(A)$, $gg^{-1}(\varphi(A)) = \varphi(A)$, whence the first part of the lemma.

Now

$$
\begin{aligned}
A \text{ is recursive} &\Longleftrightarrow g^{-1}(\varphi(A)) \text{ is recursive} \\
&\Longleftrightarrow g^{-1}(\varphi(A)) \text{ and } \mathbb{N} \setminus g^{-1}(\varphi(A)) \text{ are r.e.} \\
&\Longleftrightarrow g^{-1}(\varphi(A)) \text{ and } g^{-1}(\varphi(X) \setminus \varphi(A)) \text{ are r.e.} \\
&\Longleftrightarrow g^{-1}(\varphi(A)) \text{ and } g^{-1}(\varphi(X \setminus A)) \text{ are r.e.} \\
&\Longleftrightarrow A \text{ and } X \setminus A \text{ are r.e.} \\
&\Longleftrightarrow \varphi(A) \text{ and } \varphi(X \setminus A) \text{ are r.e. (by the first part).}
\end{aligned}
$$

Also, $\varphi(A)$ is recursive if and only if $\varphi(A)$ and $\mathbb{N} \setminus \varphi(A)$ are r.e., so to prove the lemma we need to show that

$$\varphi(X \setminus A) \text{ is r.e.} \Longleftrightarrow \mathbb{N} \setminus \varphi(A) \text{ is r.e.}$$

Since $\varphi(X \setminus A) = \varphi(X) \setminus \varphi(A) = \varphi(X) \cap (\mathbb{N} \setminus \varphi(A))$ (because φ is injective), and $\varphi(X)$ is recursive, if $\mathbb{N} \setminus \varphi(A)$ is r.e. then $\varphi(X \setminus A)$ is r.e. by Lemmas 3.2 and 3.3.

Also, $\mathbb{N} \setminus \varphi(A) = (\mathbb{N} \setminus \varphi(X)) \cup (\varphi(X) \setminus \varphi(A)) = (\mathbb{N} \setminus \varphi(X)) \cup \varphi(X \setminus A)$ and $\mathbb{N} \setminus \varphi(X)$ is r.e. by Lemma 3.2. Hence if $\varphi(X \setminus A)$ is r.e., then $\mathbb{N} \setminus \varphi(A)$ is r.e. by Lemma 3.3. $\qquad\square$

Let A be a finite set; we consider Gödel numberings of languages with alphabet A. Fix a bijection $\{1, 2, \ldots, n\} \to A$, $i \mapsto a_i$. The following can be shown to be Gödel numberings of A^*:

(1) $\varphi_1(a_{i_1} \ldots a_{i_k}) = \sum_{j=1}^{k} i_j (n+1)^{j-1}$; $\quad \varphi_1'(a_{i_1} \ldots a_{i_k}) = \sum_{j=1}^{k} i_j n^j$.

(2) $\varphi_2(a_{i_1} \ldots a_{i_k}) = 2^k \prod_{j=1}^{k} p_j^{i_j}$ (recall that p_j is the j th odd prime for $j \geq 1$).

In all three cases, the notions of r.e. and recursive subset of A^* given by Lemma 3.8 are the same, and independent of the choice of bijection $\{1, 2, \ldots, n\} \to A$ (exercise-cf Exercise 1 at the end of the chapter). Note that $\varphi_1(\varepsilon) = \varphi_1'(\varepsilon) = 0$ and $\varphi_2(\varepsilon) = 1$. Further, we can allow A to be empty ($n = 0$), when $A^* = \{\varepsilon\}$. The first assertion of the exercise is then obvious, and the second irrelevant.

Definition. A subset L of A^* is *r.e.* (resp. *recursive*) if $\varphi(L)$ is r.e. (resp. recursive), where φ can be φ_1, φ_1' or φ_2.

Note. If $B \subseteq A$ then $\varphi(B^*)$ is recursive, where φ can be φ_1, φ_1' or φ_2 (exercise). Hence $\varphi|_{B^*}$ is a Gödel numbering of B^*.

Now let $G = (V_T, V_N, P, S)$ be a grammar and put $A = V_T \cup V_N$. Fix a numbering of A, say $A = \{a_1, \ldots, a_n\}$, and use the Gödel numbering $\varphi = \varphi_2$ defined above.

Also number the productions, say $P = \{\alpha_1 \longrightarrow \beta_1, \ldots, \alpha_l \longrightarrow \beta_l\}$, and let $\lambda_i = |\alpha_i|$, $\mu_i = |\beta_i|$ (the lengths of the words).

Lemma 3.9. *For $1 \leq i \leq l$, there is a primitive recursive function $f_i : \mathbb{N}^2 \to \mathbb{N}$ such that, if $m = \varphi(x_1 \ldots x_k)$ and $x_1 \ldots x_k = x_1 \ldots x_{r-1} \alpha_i x_{r+\lambda_i} \ldots x_k$, then*

$$f_i(r,m) = \varphi(x_1 \ldots x_{r-1} \beta_i x_{r+\lambda_i} \ldots x_k)$$

and $f_i(r,m) = m$ otherwise.

Proof. First define $F(r,m,s,t) = \begin{cases} p_t^{\log p_r(m)} \cdots p_{t+s-1}^{\log p_{r+s-1}(m)} & \text{if } r,s \geq 1 \\ 1 & \text{otherwise} \end{cases}$

and show F is primitive recursive (exercise). Then use F to define f_i. It is recommended that the reader tries to do this, as the answer below is a rather complicated expression.

$$f_i(r,m) = \begin{cases} 2^{(\log_2(m)+\mu_i) \dot{-} \lambda_i} . F(1,m,r \dot{-} 1,1) . F(1,\varphi(\beta_i),\mu_i,r) \times \\ F(r+\lambda_i, m, (\log_2(m)+1) \dot{-} (r+\lambda_i), r+\mu_i), \\ \quad \text{if } 2^{\lambda_i} F(r,m,\lambda_i,1) = \varphi(\alpha_i),\ 1 \leq r \leq \log_2(m)+1 \text{ and } m \in \varphi(A^*) \\ m, \quad \text{otherwise.} \end{cases}$$

It is left to the reader to show f_i is primitive recursive; this needs the fact that "$m \in \varphi(A^*)$" is a primitive recursive predicate, which is part of Exercise 1 at the end of the chapter. □

Proposition 3.10. *A type 0 language is r.e.*

Proof. Let $L = L_G$, with G a grammar as above. With the notation of Lemma 3.9, put

$$f(u,r,m) = f_i(r,m) \quad \text{if } u \equiv i \mod l.$$

Then f is primitive recursive (exercise). Define $g : \mathbb{N}^2 \to \mathbb{N}$ by

$$g(x,0) = \varphi(S) \quad (S = \text{start symbol of } G)$$
$$g(x,t+1) = f(\log_2(x), \log_3(x), g(\log_5(x),t)).$$

Then g is primitive recursive (this follows easily from Chap. 2, Exercise 8).

Let X_j be the set of all $\alpha \in A^*$ such that $S \overset{\bullet}{\longrightarrow} \alpha$ by a derivation of length at most j. Then $\varphi(X_j) = \{g(x,j) \mid x \in \mathbb{N}\}$ (by induction on j). Let $X = \bigcup_{j \geq 0} X_j$. Then $\varphi(X) = g(\mathbb{N}^2) = g \circ J^{-1}(\mathbb{N})$ (see Chap. 2, Exercise 3), so $\varphi(X)$ is r.e. Also, $L = X \cap V_T^*$, so $\varphi(L) = \varphi(X) \cap \varphi(V_T^*)$ is r.e., since $\varphi(V_T^*)$ is recursive, and so r.e. □

Proposition 3.11. *A type 1 (context-sensitive) language is recursive.*

Proof. Let $L = L_G$ where all productions of G have the form $\alpha \longrightarrow \beta$ with $|\alpha| \leq |\beta|$. (We are assuming for the moment that $S \to \varepsilon$ is not a production.) Use the notation of Prop. 3.10. For $k \geq 1$, let $Y_j(k)$ be the set of elements in X_j of length at most k. Then

$$Y_j(k) = Y_{j-1}(k) \cup \{\alpha \in A^* \mid \exists \beta \in Y_{j-1}(k) \text{ such that } \beta \text{ rewrites to } \alpha \text{ and } |\alpha| \leq k\}.$$

The number of words of length at most k in A^+ is $n + n^2 + \ldots + n^k$ (recall: n is the cardinality of A). Abbreviating $Y_j(k)$ to Y_j, if $Y_0 \subsetneq Y_1 \subsetneq \ldots \subsetneq Y_j$ then the number of elements of Y_j is at least j, so $j \leq n + n^2 + \ldots + n^k$. Hence $Y_{j-1} = Y_j$ for some $j \leq 1 + n + n^2 + \ldots + n^k$, and then $Y_{j-1} = Y_j = Y_{j+1} = \ldots$.

Next, we claim that there is a primitive recursive function $G : \mathbb{N}^2 \to \mathbb{N}$ such that $\alpha \in Y_j(k)$ if and only if $\varphi(\alpha) = g(x, j)$ for some $x \leq G(k, j)$. (See the proof of Prop. 3.10 for the definition of g.) If $j = 0$ we can take $G(k, 0) = 0$.

Suppose $\alpha \in Y_{j+1}(k)$. If $\alpha \in Y_j(k)$, $\varphi(\alpha) = g(x, j)$ for some x, and $g(x, j) = f(0, k+2, g(x, j))$ (from the definition of f and the f_i). By definition of g, this equals $g(2^0 3^{k+2} 5^x, j+1)$.

Otherwise, β rewrites to α for some $\beta \in Y_j(k)$, so $\varphi(\beta) = g(x, j)$ for some x, and $\varphi(\alpha) = f(u, r, g(x, j))$ for some $u \leq l$ and $r \leq k+1$, which equals $g(2^u 3^r 5^x, j+1)$. Thus we can put $G(k, j+1) = 2^l 3^{k+2} 5^{G(k,j)}$, defining G by primitive recursion. Now

$$\alpha \in X \Leftrightarrow \exists j \leq (1 + n + \ldots + n^{|\alpha|})(\alpha \in Y_j(|\alpha|))$$
$$\Leftrightarrow \exists j \leq (1 + n + \ldots + n^{\log_2(\varphi(\alpha))})(\exists x \leq G(\log_2(\varphi(\alpha)), j)(\varphi(\alpha) = g(x, j))$$
$$\Leftrightarrow P(\varphi(\alpha)) \text{ where } P \text{ is a primitive recursive predicate.}$$

Therefore $\varphi(X) = \{z \in \mathbb{N} \mid P(z)\} \cap \varphi(A^*)$ is recursive, so $\varphi(L) = \varphi(X) \cap \varphi(V_T^*)$ is recursive.

Finally, if the production $S \longrightarrow \varepsilon$ is added, the new language is $L \cup \{\varepsilon\}$ (see Cor. 1.2), which is also recursive (this follows easily from Lemma 2.2).   □

Note. (1) A recursive language need not be context-sensitive. See [20, Theorem 8.3] for an example.

(2) Let $S \subset \mathbb{N}$ be r.e. and non-recursive (such sets exist by Prop. 3.6). Let $\varphi : A^* \to \mathbb{N}$ be one of the Gödel numberings defined above. There is a strictly increasing recursive function $f : \mathbb{N} \to \varphi(A^*)$, by Lemma 3.4. Then $f(S)$ is r.e. and non-recursive by Lemma 3.3 $(f^{-1}(f(S)) = S)$. Further, $\varphi(\varphi^{-1}(f(S))) = f(S)$, so $\varphi^{-1}(f(S))$ is a r.e., non-recursive language.

We shall show that a r.e. language is type 0, so there are inclusions of classes of languages:

$$\{\text{context-sensitive langs}\} \subsetneq \{\text{recursive langs}\} \subsetneq \{\text{r.e. langs}\} = \{\text{type 0 langs}\}.$$

Theorem 3.12. *For a language L, the following are equivalent.*

(1) *L is of type 0.*
(2) *L is r.e.*

(3) L *is recognised by a deterministic* TM.
(4) L *is recognised by a* TM.

Proof. By Prop. 3.10, (1) $\Rightarrow$ (2), (3) $\Rightarrow$ (4) obviously, and (4) $\Rightarrow$ (1) by Theorem 1.11. It remains to show (2) $\Rightarrow$ (3). Assume L is r.e. We can assume the alphabet of L is $A = \{2, 3, \ldots, r-1\}$ and use the Gödel numbering $\varphi : x_1 \ldots x_k \mapsto x_1 + x_2 r + \ldots + x_k r^{k-1}$ of A^*.

Consider deterministic TM's with tape alphabet $\{0, 1, 2, \ldots, r-1\}$, input alphabet $I = A$ and set of final states $F = \emptyset$. The blank symbol B will be 0.

Step 1. The following such TM's can be constructed (exercise).

R: moves right one square and halts; L: similarly.
$\widetilde{R}$: moves right until two consecutive zeros are scanned, then halts.
$\widetilde{L}$: similarly.
$P_1(i)$: prints i 1's on the tape to the right of the scanned square, starting with the scanned square, moves right one square and halts.
$P_0(i)$: similarly.
Test $\{T_0, T_1, \ldots, T_{r-1}\}$: if a is on the initially scanned square, this follows the computation of T_a. Here T_a is any TM of the form described above. (This is a simple generalisation of Example (5) of a numerical TM preceding Theorem 2.21.)

Now take

$$T_0 = \widetilde{R}L$$
$$T_1 = L$$
$$T_i = P_0(1)\widetilde{R}P_1(i)\widetilde{L}RR \quad (i \geq 2)$$

and let T be *Test* $\{T_0, T_1, \ldots, T_{r-1}\}$ with the halting states of $T_1, T_2, \ldots T_{r-1}$ identified with the initial state of *Test* $\{T_0, T_1, \ldots, T_{r-1}\}$.

Then T, started on the tape description $\underline{x}_1 \ldots x_k$ ($2 \leq x_i \leq r-1$, $k \geq 0$), halts with tape description $01^{x_1}0 \ldots 01^{x_k}\underline{0}$. Further, the halting state of T_0 is a halting state for T.

Step 2. By Exercise 14 in Chap. 2, there is a numerical TM T' which, started on tape description $01^{x_1}0 \ldots 01^{x_k}\underline{0}$, halts with tape description $\underline{0}1^{\varphi(x_1 \ldots x_k)}$, and T' has a halting state.

Step 3. There is a partial recursive function f such that $\varphi(L) = \mathrm{dom}(f)$. By Cor. 2.22, there is a numerical TM T'' which, started on the tape description $\underline{0}1^x$, halts if and only if $f(x)$ is defined (in which case, it halts with the tape description $\underline{0}1^{f(x)}$). Further, T'' has a halting state, say h.

Now $TT'T''$, started on the tape description $\underline{x}_1 \ldots x_k$ ($2 \leq x_i \leq r-1$), halts if and only if $x_1 \ldots x_k \in L$. Modify this TM by letting the set of final states be $\{h\}$, to get a deterministic TM recognising L. $\square$

Theorem 3.13. *For a language with alphabet A, the following are equivalent.*

(1) *L is recursive;*
(2) *L is recognised by a deterministic TM which, on input $x_1 \ldots x_k$ ($x_1, \ldots, x_k \in A$), always halts.*

Proof. Assume L is recursive. Again we can assume $A = \{2, 3, \ldots, r-1\}$. In the construction of Theorem 3.12, replace T'' by a TM which computes $\chi_{\varphi(L)}$, to obtain a TM U which, started on tape description $\underline{x}_1 \ldots x_k$ (where $w = x_1 \ldots x_k \in A^*$) halts,

with tape description $\begin{cases} \underline{0}1 & \text{if } w \in L \\ \underline{0} & \text{(blank tape) if } w \notin L \end{cases}$

Let $U' = UR\,Test$ (see Example (4) of a numerical TM preceding Theorem 2.21).

Then U' halts $\begin{cases} \text{in a state } p_1 \text{ if } w \in L \\ \text{in a state } p_0 \text{ if } w \notin L \end{cases}$.

Modify U' by letting the the set of final states be $\{p_1\}$, to get the desired TM.

Conversely, assume (2). We can assume the TM in (2) halts whenever a final state is reached (see Remark 1.2). Let Q be the set of states and F the set of final states of the TM. Modify the TM as follows. Add a new state h. For each pair (q, a), where $q \in Q$ and a is in the tape alphabet, such that $q \notin F$ and no transition starts with qa, add a transition $qahaR$. Then replace F by $\{h\}$. The new machine recognises $A^* \setminus L$, hence L and $A^* \setminus L$ are r.e. by Theorem 3.12. Then L is recursive by Lemma 3.8. $\square$

We use Theorem 3.12 to prove a result needed in Chapter 5. The Kleene star operation is defined in Chapter 1, before Lemma 1.5.

Lemma 3.14. *If L is a r.e. language, then so is L^*.*

Proof. By Theorem 3.12, $L = L_G$ for some type 0 grammar $G = (V_N, V_T, P, S)$. By Lemma A.1 in Appendix A, we can assume all productions in P are either of the form $\alpha \longrightarrow \beta$ where $\alpha, \beta \in V_N^*$, or of the form $A \to a$, where $A \in V_N$, $a \in V_T$. Take two new symbols S', S'' not in $V_N \cup V_T$. Let G' be the grammar (V_N', V_T, P', S'), where $V_N' = V_N \cup \{S', S''\}$ and

$$P' = P \cup \{S' \longrightarrow \varepsilon,\ S' \longrightarrow S,\ S' \longrightarrow SS''\} \cup \{aS'' \longrightarrow aS,\ aS'' \longrightarrow aSS'' \mid a \in V_T\}$$

It is left to the reader to check that $L_{G'} = L^*$, so again by Theorem 3.12, L^* is r.e. $\square$

Complexity. Turing machines are intended as models of computation. For a discussion of how a modern computer can be simulated by a deterministic TM (and vice-versa), see §8.6 in [22]. It is convenient here to use a multi-tape Turing machine, one of several variants of TM's discussed, for example, in [20], §§6.5 and 6.6, or in [21], §§7.6 and 7.8. Given a problem with an algorithm to solve it which can be implemented by a computer program, it is important to know how much time the program takes to run, in terms of some measure of complexity of its input.

In terms of a TM, the input is a word on the input tape, and we can take the length of the word as a measure of complexity. The time taken to run is measured by the number of moves the machine makes with a given input. (See [21, §12.1] for further details.) If, for any input word of length n, a TM makes at most $f(n)$

moves before halting, T is said to have *time complexity* $f(n)$. It is assumed in [21] that the TM always reads its entire input and verifies it has been read by reading a blank cell, so for input of length n, always makes at least $n + 1$ moves. Thus, for an arbitrary function f, time complexity $f(n)$ actually means time complexity $\max(n + 1, \lceil f(n) \rceil)$.

A language L is said to be in the class $\text{NTIME}(f(n))$ if $L = L(T)$ for some TM of time complexity $f(n)$. It is said to be in $\text{DTIME}((f(n))$ if $L = L(T)$ for some deterministic TM of time complexity $f(n)$. It can be shown that if $L \in \text{NTIME}(f(n))$, then $L = L(T)$ for some single tape TM T of time complexity $f(n)^2$, and if $L \in \text{DTIME}(f(n))$, T can be taken to be deterministic. See [21], Theorem 12.5 and its corollary.

We define $\mathcal{NP}$ to be the class of languages which are in $\text{NTIME}(f(n))$ for some polynomial $f(n)$. Also, $\mathcal{P}$ is the class of languages in $\text{DTIME}(f(n))$ for some polynomial $f(n)$. It is believed that the class of problems which can be efficiently solved by a computer are those having an algorithm implemented by a program which runs in polynomial time. It is argued in [22], §8.6.3, that a deterministic TM simulating such a computer program is of polynomial time complexity. This explains the interest in the class $\mathcal{P}$.

Clearly $\mathcal{P} \subseteq \mathcal{NP}$, but by contrast with Theorem 3.12, it is unknown whether or not $\mathcal{P} = \mathcal{NP}$, indeed this has become a notorious problem. For examples of some problems which might lead to languages in $\mathcal{NP} \setminus \mathcal{P}$, and the related idea of an NP-complete language, see [22, Chap. 10].

One can also consider space bounds for computations. In a TM, the space used is measured by the maximum number of cells scanned on each tape. This leads to language classes $\text{NSPACE}(f(n))$ and $\text{DSPACE}(f(n))$. For further information, see [21, §12.1].

Exercises on Chapter 3

1. Consider the functions φ_1, φ_1' and φ_2 defined after Lemma 3.8.

   (a) Show that φ_1 and φ_2 are Gödel numberings, and that φ_1 and φ_2 give the same collections of r.e. and recursive subsets of A^*.
   (b) Show that φ_1' is a Gödel numbering, and that φ_1' and φ_1 give the same collections of r.e. and recursive subsets of A^*.
   (c) Show that the collections of r.e. and recursive subsets given by φ_1, φ_1' and φ_2 do not depend on the choice of bijection $\{1, 2, \ldots, n\} \to A$.
   (d) Show that if $B \subseteq A$, then $\varphi(B^*)$ is recursive, where φ is either φ_1, φ_1' or φ_2.

   (Warning: this is quite tricky.)
2. Construct the TM s used in Step 1 of the proof of Theorem 3.12.
3. Show that, if L and L' are r.e. languages, then LL' is r.e.

Chapter 4
Context-free Languages

In this chapter we study context-free languages and the machines recognising them, the pushdown stack automata. The class of languages recognised by deterministic pushdown stack automata is called the class of *deterministic* languages. It is a proper subclass of the class of context-free languages. The class of deterministic languages is the class of languages generated by what are called $LR(k)$ grammars (k being a natural number). However, things are complicated by the fact that a pushdown stack automaton has two ways of recognising a language. In the case of deterministic machines, this makes a difference to the class of languages recognised, leading to a proper subclass of the deterministic languages. It turns out that this class is precisely the class of languages generated by $LR(0)$ grammars. The idea of $LR(k)$ language is important in Computer Science in the construction of parsers, although our account does not reflect this. (See, for example, the parser-generator YACC described in [22, §5.3.2].)

Subsequently, in dealing with grammars, we shall just say *terminal* instead of terminal symbol, and a non-terminal symbol will be called a *variable*. Also, *terminal string* means a word whose letters are all terminals. As a matter of notation, if G is a grammar with set of productions P, we shall sometimes write $\alpha \xrightarrow[G]{\bullet} \beta$ to mean $\alpha \xrightarrow[P]{\bullet} \beta$, and refer to a P-derivation as a G-derivation. We also write $\alpha \xrightarrow[G]{} \beta$ to mean α rewrites to β using a production of P.

It is convenient to extend the definition of context-free grammar. A context-free grammar with ε-productions is a grammar in which all productions are either context-free or *ε-productions*, that is, productions of the form $A \rightarrow \varepsilon$, where A is a variable. Hitherto, the only such production allowed is $S \rightarrow \varepsilon$, where S is the start symbol. We begin by showing that allowing ε-productions does not change the class of languages generated.

Lemma 4.1. *Let $G = (V_N, V_T, P, S)$ be a context-free grammar with ε-productions. Then there is a context-free grammar G_1 with $L_G = L_{G_1}$. Further, we can assume the start symbol of G_1 does not appear on the right-hand side of any production.*

I. Chiswell, *A Course in Formal Languages, Automata and Groups*,
DOI 10.1007/978-1-84800-940-0_4,
© Springer-Verlag London Limited 2009

Proof. By Lemma 1.1, we can assume S does not occur on the right-hand side of any production in P. Let $\mathcal{N}$ be the set of variables A in V_N such that $A \xrightarrow{\bullet}_{G} \varepsilon$. (The set $\mathcal{N}$ can be found by the following procedure, starting with $\mathcal{N}$ empty.

(1) If $A \to \varepsilon$ is a production, then put A in $\mathcal{N}$.
(2) If $A \to B_1 \ldots B_k$ is a production and $B_i \in \mathcal{N}$ for $1 \le i \le k$, then put A in $\mathcal{N}$.
(3) Repeat step (2) until no more variables are put in $\mathcal{N}$.

The proof that this finds all elements of $\mathcal{N}$ is left to the reader. An induction on the number of steps in a derivation from A to ε, for $A \in \mathcal{N}$, is required.)

We now modify the productions of G as follows.

(1) Remove all ε-productions.
(2) If $A \longrightarrow X_1 \ldots X_r$ is in P, where $X_i \in V_T \cup V_N$ and $r > 0$, replace this by all productions of the form $A \to Y_1 \ldots Y_r$, where, if $X_i \in \mathcal{N}$, Y_i is either X_i or ε, otherwise Y_i is X_i. This replaces the production by 2^m productions (including the original production), where m is the number of symbols X_i in $\mathcal{N}$. However, if all $X_i \in \mathcal{N}$, $A \longrightarrow \varepsilon$ is omitted.

Call the new set of productions P' and let $G' = (V_N, V_T, P', S)$. We claim that $L_{G'} = L_G \setminus \{\varepsilon\}$. Since G' is context-free and contains no ε-productions, $\varepsilon \notin L_{G'}$. In a G'-derivation, any use of a production $A \to Y_1 \ldots Y_r$ as in (2) can be replaced by use of the production $A \longrightarrow X_1 \ldots X_r$, followed by several uses of ε-productions of G, to obtain a G-derivation of the same word. Hence $L_{G'} \subseteq L_G \setminus \{\varepsilon\}$. To prove equality, we show that, for $A \in V_N$

$$A \xrightarrow{\bullet}_{G} w \text{ and } w \ne \varepsilon \text{ implies } A \xrightarrow{\bullet}_{G'} w.$$

The proof is by induction on the number of steps in a G-derivation from A to w. If the number is 1, then $A \longrightarrow w$ is a production of G, so of G' since $w \ne \varepsilon$. If the number of steps is $k > 1$, the derivation has the form $A, X_1 \ldots X_n, \ldots, w$, where $A \longrightarrow X_1 \ldots X_n$ is in P. We can write $w = w_1 \ldots w_n$, where $X_i \xrightarrow{\bullet}_{G} w_i$ by a derivation of length less than k, so by induction $X_i \xrightarrow{\bullet}_{G'} w_i$, if $w_i \ne \varepsilon$. Let $Z_1, \ldots Z_m$ be those X_i (in order) for which $w_i \ne \varepsilon$. Note that $m > 0$ since $w \ne \varepsilon$. Then $Z_1 \ldots Z_m \xrightarrow{\bullet}_{G'} w$, and $A \longrightarrow Z_1 \ldots Z_m$ is in P', so $A \xrightarrow{\bullet}_{G'} w$.

If $\varepsilon \in L_G$ (i.e. $S \in \mathcal{N}$), we let G_1 be G' with $S \longrightarrow \varepsilon$ added to the productions, and $G_1 = G'$ otherwise. Then G_1 is context-free, S does not appear on the right-hand side of any production of G_1, and $L_G = L_{G_1}$ (see the proof of Cor. 1.2). $\square$

For the rest of this chapter, "context-free grammar" will mean a context-free grammar with ε-productions. A useful idea in dealing with context-free grammars is that of a parsing tree. A rooted tree is a tree with a distinguished vertex, v_0, called the root. This establishes a *level* for each vertex v, namely the length (number of edges in) of the reduced path from v_0 to v. Then

(1) v_0 is the only vertex of level 0

(2) every vertex v of level $n > 0$ is adjacent to exactly one vertex of level $n-1$, and
    v is called a *successor* of this vertex.

A vertex with no successors is called a *leaf*. A vertex v is a *descendant* of a vertex
w if there is a sequence of vertices $w = v_0, v_1, \ldots, v_n = v$ for some $n \geq 0$, such that
v_i is a successor of v_{i-1} for $1 \leq i \leq n$.

   We shall consider only finite rooted trees, which can be drawn in the plane with
the root at the top and vertices of the same level physically at the same level. This
gives extra structure to the tree; the successors of a given vertex are linearly ordered
"from left to right". If v, w are two successors of the same vertex with v to the left
of w, then all descendants of v are said to be to the left of all descendants of w. It is
an exercise to show this induces a linear ordering on the leaves.

Definition. Let $G = (V_N, V_T, P, S)$ be a context-free grammar. Let $A \in V_N$. An *A-tree*
for G is a finite rooted tree whose vertices are labelled by elements of $V_N \cup V_T \cup \{\varepsilon\}$,
satisfying the following.

(1) the label on the root is A.
(2) if a vertex is a non-leaf, its label is in V_N.
(3) if a non-leaf has label B, and the successors of this vertex have labels $X_1, \ldots, X_n$
     in order from left to right, then $B \longrightarrow X_1 \ldots X_n$ is in P.
(4) if a leaf v is a successor of w and has label ε, it is the only successor of w.

An A-tree for some variable A is called a *parsing tree* of G.

   A *subtree* of a parsing tree is a non-leaf of the tree together with all its descendants, the edges joining them, their labels and left-right ordering. If B is the label on
the vertex, then this is a *B-tree*.

Definition. The *yield* of a parsing tree is the word obtained by reading the labels on
the leaves from left to right.

Example. Let $G = (\{S, A, B\}, \{a, b, c\}, P, S)$ where P contains

$$S \longrightarrow Bc, \quad B \longrightarrow aAb, \quad A \longrightarrow aAb, \quad A \longrightarrow ab.$$

Here is an S-tree for G and a subtree which is a B-tree (we just indicate the vertices
by their labels):

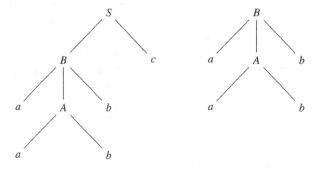

Figure 4.1

The yield of the parsing tree is a^2b^2c and of the subtree a^2b^2. Notice that the vertex of the parsing tree labelled a at level 2 is to the left of the vertex labelled a at level 3, according to our definitions, although not physically so.

Definition. A *leftmost derivation* is one in which, at each step, the production used replaces the leftmost variable. Similarly for rightmost.

Lemma 4.2. *Let* $\alpha \in (V_N \cup V_T)^*$, $A \in V_N$.

(1) $A \overset{\bullet}{\longrightarrow} \alpha$ *if and only if there is an A-tree with yield* α.
(2) *If* $\alpha \in V_T^*$, *and there is an A-tree with yield* α, *then there is a leftmost derivation of* α *from A, and a rightmost derivation of* α *from A.*

Proof. (1) Given a derivation $A = \alpha_0, \alpha_1, \ldots, \alpha_n = \alpha$, we construct inductively A-trees $R_0, R_1, \ldots R_n$ such that R_i has yield α_i. We take R_0 to consist of a single vertex (which is both the root and a leaf) with label A. Suppose R_{i-1} has been defined. Let $B \longrightarrow \beta$ be the production used to get from α_{i-1} to α_i, and let $\beta = X_1 \ldots X_k$, where $X_i \in V_N \cup V_T$. Then there is an occurrence of B in α_{i-1} which is replaced by β, and this corresponds to a leaf of R_{i-1}. Add k successors to this leaf, with labels $X_1, \ldots, X_k$ (in left-right order), to obtain R_i.

The R_i satisfy: R_i is obtained from R_{i-1} by adding successors to a leaf corresponding to a production $B \longrightarrow \beta$;
R_0 consists of a single vertex with label A.

Conversely, suppose R is an A-tree with yield α. We shall construct a sequence of A-trees R_i with $R_n = R$ for some n. This sequence will have the additional property that, if v is a non-leaf of R_i, then R_i contains all the successors of v in R. We take R_0 to be the root v_0 of R, with label A. Suppose R_{i-1} has been defined. Choose a leaf of R_{i-1} which has successors in R, and add all these with their labels, to obtain R_i. Suppose no such leaf exists; then we claim $R_{i-1} = R$ and we put $n = i - 1$. For if v is a vertex of R not in R_{i-1}, let $v_0, v_1, \ldots, v_r = v$ be the vertices, in order, of the reduced path from v_0 to v in R_{i-1}. Let j be smallest such that v_j is not in R_{i-1}. Then $j > 0$ since v_0 is in R_{i-1}, and v_{j-1} is in R_{i-1}. Now v_{j-1} has v_j as a successor in R, so by assumption is not a leaf of R_{i-1}. Hence R_{i-1} contains all successors of v_{j-1} in R, including v_j, a contradiction.

Let α_i be the yield of R_i. Then it is easily seen by induction on i that $A = \alpha_0, \alpha_1, \ldots, \alpha_i$ is a derivation, and taking $i = n$ gives a derivation of α.

(2) The procedure just given involves choices of leaves, so in general will give several different derivations. To make it unique, always choose the leftmost possible leaf (in the left-right order of the leaves of R_{i-1}. If $\alpha \in V_T^*$, this will give a leftmost derivation of α. (All leaves to the left of the one chosen must be leaves of R, and therefore have labels in V_T, so the leaf chosen corresponds to the leftmost variable in α_{i-1}.) Similarly, if at each stage we choose the rightmost leaf, we get a rightmost derivation. $\qquad \square$

Remark 4.1. In the proof of Lemma 4.2, suppose R, R' are two different A-trees. (Here, "different" means "non-isomorphic", where two parsing trees are isomorphic if they are isomorphic as trees, via an isomorphism preserving roots, labels and left-right orderings. That is, they look exactly the same when drawn.) Let $R_0, R_1, \ldots$ and $R'_0, R'_1, \ldots$ be the sequences of A-trees constructed from them, always choosing the leftmost leaf. Let $\alpha_0, \alpha_1, \ldots$ and $\alpha'_0, \alpha'_1, \ldots$ be the corresponding derivations. Then there is a first value of i such that R_i and R'_i are different. Then $\alpha_j = \alpha'_j$ for $j < i$, but $\alpha_i \neq \alpha'_i$. These are therefore two distinct derivations. Similarly, taking the sequences obtained by always choosing the rightmost leaf gives two different derivations.

If $A = \alpha_0, \alpha_1, \ldots$ and $A = \alpha'_0, \alpha'_1, \ldots$ are two leftmost derivations, construct the sequences of A-trees $R_0, R_1, \ldots, R_n = R$ and $R'_0, R'_1, \ldots, R'_m = R'$ as in the proof of Lemma 4.2. Then these are the sequences obtained from R and R' by always choosing the leftmost possible leaf (by an inductive proof). Therefore, if R and R' are the same, these sequences of A-trees will be the same (again by an inductive argument-isomorphisms preserve left-right ordering). Thus if the two derivations are different, the sequences of A-trees will be different, at the point where the derivations first differ, so R and R' will be different.

Similarly, two different rightmost derivations give two different parsing trees.

Definition. A context-free grammar is *ambiguous* if there exists $w \in V_T^*$ and two different S-trees with yield w.

In view of Remark 4.1, this is equivalent to saying there exists $w \in V_T^*$ having two different leftmost derivations from S, also to saying there exists $w \in V_T^*$ having two different rightmost derivations from S.

We now consider ways of modifying context-free grammars so they generate the same language, but have certain extra properties, culminating in two normal forms.

Definition. Let $G = (V_N, V_T, P, S)$ be a grammar.

(1) A letter $X \in V_N \cup V_T$ is called *generating* if $X \xrightarrow[G]{\bullet} w$ for some $w \in V_T^*$.

(2) A letter $X \in V_N \cup V_T$ is called *reachable* if $S \xrightarrow[G]{\bullet} \alpha X \beta$ for some $\alpha, \beta \in (V_N \cup V_T)^*$.

Note that every element of V_T is generating.

Lemma 4.3. *Let $G = (V_N, V_T, P, S)$ be a context-free grammar with $L_G \neq \emptyset$. There is a context-free grammar $G' = (V'_N, V_T, P', S)$, such that every $A \in V'_N$ is generating, with $L_G = L_{G'}$.*

Proof. Let V'_N be the set of all $A \in V_N$ which are generating and P' the set of productions in P having all their letters in $V'_N \cup V_T$. (The set $\mathscr{G}$ of all generating symbols can be found by the following procedure, starting with $\mathscr{G} = V_T$.

(1) If $A \to \alpha$ is a production, and every letter of α is in $\mathscr{G}$, then add A to $\mathscr{G}$;
(2) Repeat step (1) until no new letters are added to $\mathscr{G}$.

The proof that this works is left to the reader. Then of course, $V'_N = \mathscr{G} \cap V_N$.)

Note that $S \in V'_N$ by the assumption $L_G \neq \emptyset$. Clearly $L_{G'} \subseteq L_G$. Suppose $w \in L_G$, $w \notin L_{G'}$. There is a G-derivation of w from S, which uses a production not in G', so some word in the derivation has the form $w_1 A w_2$, where $A \notin V_N'$. Since $w_1 A w_2 \xrightarrow[G]{\bullet} w$, $A \xrightarrow[G]{\bullet} w'$ for some $w' \in V_T^*$, so A is generating, a contradiction. This gives the desired grammar.   □

Lemma 4.4. *Let $G = (V_N, V_T, P, S)$ be a context-free grammar. There is a context-free grammar $G' = (V_N', V_T', P', S)$, such that every $A \in V_N' \cup V_T'$ is reachable, with $L_G = L_{G'}$.*

Proof. Let V_N' be the set of all reachable letters in V_N and V_T' the set of all reachable letters in V_T. (We can find the set $\mathscr{R}$ of all reachable letters, hence V_N' and V_T', as follows.

(1) Start with $\mathscr{R} = \{S\}$.
(2) If $A \longrightarrow \alpha$ is a production and $A \in \mathscr{R}$, add all letters occurring in α to $\mathscr{R}$.
(3) Repeat step (2) until no more letters are added to $\mathscr{R}$.

The proof that this works is left to the reader.)

Now let P' be the set of productions in P having all their letters in $V_N' \cup V_T'$. Clearly $L_{G'} \subseteq L_G$. But in a G-derivation from S, all letters which occur are reachable, so all productions used are in P', hence it is a G'-derivation, so $L_{G'} = L_G$.   □

Definition. A letter in a grammar G is called *useless*, or a *useless symbol*, if it does not appear in any derivation of an element of V_T^* from S, otherwise it is called *useful*.

It is left as an easy exercise to show that a letter is useful if and only if it is both generating and reachable.

Lemma 4.5. *Every non-empty context-free language is generated by a grammar with no useless symbols.*

Proof. Let $L = L_G$ where G is context-free. Let G_1 be the grammar obtained from G by Lemma 4.3, with all letters generating, and let G_2 be the grammar obtained from G_1 by Lemma 4.4, with all letters reachable, so $L_G = L_{G_2}$. Suppose G_2 has a useless symbol X. Then X is reachable, so $S \xrightarrow[G_2]{\bullet} \alpha X \beta$ for some α, β. Since the productions of G_2 are productions of G_1, it follows that $S \xrightarrow[G_1]{\bullet} \alpha X \beta \xrightarrow[G_1]{\bullet} w$ for some terminal string w. But then all letters in this derivation are reachable, so this is a G_2-derivation and X is not useless, a contradiction.   □

Lemma 4.6. *If L is a context-free language, then $L = L_G$ for some context-free grammar G having no productions of the form $A \longrightarrow B$, where A and B are variables.*

Proof. Suppose $L = L_{G'}$ where $G' = (V_N, V_T, P, S)$ is a context-free grammar. Let $\mathscr{U}$ be the set of all ordered pairs (A, B), where $A, B \in V_N$, such that $A \xrightarrow[G]{\bullet} B$. (The set $\mathscr{U}$ can be found by the following procedure.

(1) Start with $\mathscr{U} = \{(A, A) \mid A \in V_N\}$.

(2) If $(A,B) \in \mathcal{U}$ and $B \longrightarrow C$ is a production, where $C \in V_N$, then add (A,C) to $\mathcal{U}$.

(3) Repeat step (2) until no more pairs are added to $\mathcal{U}$.

The proof that this works is left to the reader.)

Define a new set of productions R as follows: for each $(A,B) \in \mathcal{U}$, R contains all productions $A \longrightarrow \alpha$, where $B \longrightarrow \alpha$ is a production in P with $\alpha \notin V_N$. (Note that R contains all productions in P of the form $A \longrightarrow \alpha$ with $\alpha \notin V_N$, as $(A,A) \in \mathcal{U}$.) Now let $G = (V_N, V_T, R, S)$. Clearly if $A \longrightarrow \alpha$ is in R, then $A \xrightarrow[G]{\bullet} \alpha$, so $L_{G'} \subseteq L_G$.

Suppose $w \in L_G$ and consider a leftmost derivation

$$S = \alpha_0 \xrightarrow[G]{} \alpha_1 \xrightarrow[G]{} \ldots \xrightarrow[G]{} \alpha_n = w.$$

Suppose there is a sequence $\alpha_i \xrightarrow[G]{} \alpha_{i+1} \xrightarrow[G]{} \ldots \xrightarrow[G]{} \alpha_j$ using only productions of the form $A \longrightarrow B$, but $\alpha_j \xrightarrow[G]{} \alpha_{j+1}$ by a production in R. (We cannot have $j = n$ since $w \in V_T^*$.) Then $\alpha_i, \ldots, \alpha_j$ all have the same length, and since the derivation is leftmost, the letter replaced at each stage must be in the same position. If the letter replaced in α_i is A and the letter in the same position in α_j is B, then $A \xrightarrow[G]{\bullet} B$, and $\alpha_j \xrightarrow[G]{} \alpha_{j+1}$ by a production $B \longrightarrow \beta$. But then $\alpha_i \xrightarrow[G']{} \alpha_{j+1}$ by the production $A \longrightarrow \beta$ of G'. Thus we can remove $\alpha_{i+1}, \ldots, \alpha_j$ from any such sequence to obtain a G'-derivation of w from S. Hence $L_G = L_{G'}$. $\qquad \square$

Normal Forms. We show that a context-free language can be defined by a context-free grammar in normal form, that is, where the productions all have a certain form. There are two such normal forms, and we can now establish the first of these. It is a refined version of Lemma A.1, Appendix A, for type 2 grammars.

Theorem 4.7. (Chomsky Normal Form) *Any context-free language L with $\varepsilon \notin L$ can be generated by a grammar in which all productions are of the form $A \longrightarrow BC$ or $A \longrightarrow a$, where A, B, C are variables and a is a terminal.*

Proof. By Lemma 4.1 and the fact that $\varepsilon \notin L$, we can assume that $L = L_G$ for some context-free grammar G with no ε-productions. The construction of Lemma 4.6 does not introduce any ε-productions, so we can further assume that G has no productions of the form $A \longrightarrow B$ where A, B are variables. Then if the right-hand side of a production has a single letter, it must be a terminal, so is in the required form.

If a terminal a appears on the right in a production $A \longrightarrow X_1 \ldots X_n$, where $n > 1$, add a new variable C_a and a production $C_a \longrightarrow a$. Then replace all occurrences of a on the right of such productions by C_a. Do this for every terminal, and call the resulting grammar G'. If $\alpha \longrightarrow \beta$ is a G-production then clearly $\alpha \xrightarrow[G']{\bullet} \beta$, hence $L_G \subseteq L_{G'}$. We show by induction on the number s of steps in a derivation that if A is a variable of G and w is a terminal string of G such that $A \xrightarrow[G']{\bullet} w$, then $A \xrightarrow[G]{\bullet} w$. It then follows that $L_G = L_{G'}$.

If $s = 1$, then $A \longrightarrow w$ is a production of both G and G'. If $s > 1$, the derivation has the form $A, Y_1 \ldots Y_n, \ldots, w$, where Y_i are variables of G' and $n > 1$. Then we can write $w = w_1 \ldots w_n$, where $Y_i \xrightarrow[G']{\bullet} w_i$ by a derivation of length less than s (using some but not all of the productions used in the original derivation). If $Y_i = C_a$ for some a then the only production in this derivation is $C_a \longrightarrow a$, as this is the only one with C_a on the left-hand side, hence $w_i = a$. If Y_i is a variable of G, then by induction $Y_i \xrightarrow[G]{\bullet} w_i$.

Now the first production used in the G'-derivation is $A \longrightarrow Y_1 \ldots Y_n$, arising from a G-production $A \longrightarrow X_1 \ldots X_n$, where $Y_i = X_i$ if X_i is a variable of G, and $Y_i = C_a$ if $X_i = a$ is a terminal, in which case $w_i = a$. It follows that $X_i \xrightarrow[G]{\bullet} w_i$ for all i, hence

$$X_1 \ldots X_n \xrightarrow[G]{\bullet} w_1 \ldots w_n = w$$

and so $A \xrightarrow[G]{\bullet} w$, finishing the inductive proof.

The productions of G' are of the form $A \longrightarrow a$ and $A \longrightarrow X_1 \ldots X_n$ $(n \geq 2)$ where all X_i are variables. For a production of the second form with $n \geq 3$, we add new variables $D_1, \ldots, D_{n-2}$ and replace this production by the productions

$$A \longrightarrow X_1 D_1, \ D_1 \longrightarrow X_2 D_2, \ldots, D_{n-3} \longrightarrow X_{n-2} D_{n-2}, \ D_{n-2} \longrightarrow X_{n-1} X_n.$$

This gives a new grammar G'' in the required form. The proof that $L_{G'} = L_{G''}$ is left as an exercise.   □

For our second normal form, two lemmas are needed, giving more ways of manipulating context-free grammars while not changing the language generated. First, we introduce some notation. An A-*production* is one of the form $A \longrightarrow \alpha$. A list of A-productions $A \longrightarrow \alpha_1, \ldots, A \longrightarrow \alpha_n$ is abbreviated to $A \longrightarrow \alpha_1 | \alpha_2 | \ldots | \alpha_n$.

Lemma 4.8. *Let $G = (V_N, V_T, P, S)$ be a context-free grammar. Let $A \longrightarrow \alpha B \gamma$ be in P and let the B-productions in P be $B \longrightarrow \beta_1 | \beta_2 | \ldots | \beta_n$. Let $G' = (V_N, V_T, P', S)$ be obtained by deleting the production $A \longrightarrow \alpha B \gamma$ from P and adding the productions*

$$A \longrightarrow \alpha \beta_1 \gamma | \alpha \beta_2 \gamma | \ldots | \alpha \beta_n \gamma.$$

Then $L_G = L_{G'}$.

Proof. If $A \longrightarrow \alpha \beta_i \gamma$ is used in a step of a G'-derivation, then it can be replaced by two steps using the productions $A \longrightarrow \alpha B \gamma$ and $B \longrightarrow \beta_i$, to obtain a G-derivation, hence $L_{G'} \subseteq L_G$. If $A \longrightarrow \alpha B \gamma$ is used in a step of a G-derivation of a terminal string w, the variable B must be changed at some later step using a production $B \longrightarrow \beta_i$. These two steps can be replaced (at the point where $A \longrightarrow \alpha B \gamma$ is used) by $A \longrightarrow \alpha \beta_i \gamma$, resulting in a G-derivation of w. Hence $L_G = L_{G'}$.   □

Lemma 4.9. *Let $G = (V_N, V_T, P, S)$ be a context-free grammar, $A \in V_N$. Suppose*

$$A \longrightarrow A\alpha_1 | A\alpha_2 | \ldots | A\alpha_m$$

are the A-productions whose right-hand side begins with A, and let the other A-productions in P be $A \longrightarrow \beta_1 | \beta_2 | \ldots | \beta_n$. Add a new variable B, and let $G' = (V_N \cup \{B\}, V_T, P', S)$, where P' is obtained by replacing all the A-productions by

$A \longrightarrow \beta_1 | \beta_2 | \dots | \beta_n | \beta_1 B | \beta_2 B | \dots | \beta_n B, \text{ and } B \longrightarrow \alpha_1 | \alpha_2 | \dots | \alpha_m | \alpha_1 B | \alpha_2 B | \dots | \alpha_m B.$

Then $L_G = L_{G'}$.

Proof. If $w \in L_G$, there is a leftmost G-derivation of w from S, by Lemma 4.2. If a derivation $A \longrightarrow A\alpha_i$ is used, it must be the start of a succession of steps using a sequence of productions of the form

$$A \longrightarrow A\alpha_{i_1}, A \longrightarrow A\alpha_{i_2}, \dots, A \longrightarrow A\alpha_{i_r}, A \longrightarrow \beta_j \qquad (*)$$

resulting in A being replaced by the string $\beta_j \alpha_{i_r} \alpha_{i_{r-1}} \dots \alpha_{i_1}$. This can be replaced by steps using the sequence of productions:

$$A \longrightarrow \beta_j B, B \longrightarrow \alpha_{i_r} B, B \longrightarrow \alpha_{i_{r-1}} B, \dots, B \to \alpha_{i_2} B, B \longrightarrow \alpha_{i_1} \qquad (**)$$

The result is a G'-derivation of w from S, so $L_G \subseteq L_{G'}$. Conversely, if $w \in L_{G'}$, we can find a rightmost G'-derivation of w from S by Lemma 4.2. If B appears in this derivation, there is a succession of steps corresponding to a sequence of the form $(**)$, which can be replaced by the sequence $(*)$, resulting in a G-derivation of w from S. Hence $L_G = L_{G'}$.   □

Theorem 4.10. (Greibach Normal Form) *Every context-free language L without ε is generated by a grammar in which all productions are of the form $A \longrightarrow a\alpha$, where A is a variable, a is a terminal and α is a string of variables.*

Proof. Let $G = (V_N, V_T, P, S)$ be a grammar in Chomsky normal form generating L. Number the variables, say $V_N = \{A_1, \dots, A_n\}$, and add new variables $\{B_1, \dots, B_n\}$ (this does not change the language generated). We begin by modifying the productions so that if $A_i \longrightarrow A_j \gamma$ is a production, then $i < j$. Further, the right-hand side of a production is either a non-empty string of variables, or begins with a terminal, followed by a string of variables. If $i = 1$ we replace the A_1-productions with right-hand sides starting with A_1 using Lemma 4.9 (with $A = A_1$, $B = B_1$) to obtain the desired conditions.

Assume we have achieved the desired conditions for $1 \le i \le k$. For each production $A_{k+1} \longrightarrow A_j \gamma$ with $j \le k$, apply Lemma 4.8 to this production with $\alpha = \varepsilon$, $A = A_{k+1}$ and $B = A_j$. This replaces each such production by productions of the form $A_{k+1} \longrightarrow A_{j'} \gamma'$, where $j' > j$ and γ' is a string of variables, or $A_{k+1} \longrightarrow a\gamma'$ where a is a terminal and γ' is a string of variables. Applying this procedure at most k times brings the A_{k+1}-productions to the required form, except for productions of the form $A_{k+1} \longrightarrow A_{k+1} \gamma$. We replace each of these productions by new productions using Lemma 4.9 (with $A = A_{k+1}$, $B = B_{k+1}$) to get all A_{k+1}-productions in the required form. (If there are no productions of the form $A_{k+1} \longrightarrow A_{k+1} \gamma$, the variable B_{k+1} can be omitted.) By induction on k, we obtain the desired conditions.

Looking at the form of the B-productions in Lemma 4.9, the productions which are not in the form given by the theorem are now of two kinds.

(1) $A_i \longrightarrow A_j \gamma$ where $i < j$ and γ is a string of variables.
(2) $B_i \longrightarrow A_j \gamma$ where γ is a string of variables.

The right-hand sides of the A_n-productions must already be in the required form (terminal followed by a string of variables). The right-hand sides of the A_{n-1}-productions of type (1) start with A_n, and can be modified using Lemma 4.8 (with $A = A_{n-1}$, $B = A_n$, $\alpha = \varepsilon$) to bring them to the required form. We can continue to use Lemma 4.8 to successively bring the A_i-productions, for $i = n-2, \ldots, 1$ to the required form.

Finally all productions of type (2) can now be modified by use of Lemma 4.8 to bring them to the required form.   □

Before proceeding, we shall prove the Pumping Lemma (Lemma 1.9), whose statement we recall.

Let L be a context-free language. Then there is an integer $p > 0$, depending only on L, such that, if $z \in L$ and $|z| \geq p$, then z can be written as $z = uvwxy$, where $|vwx| \leq p$, v and x are not both ε and for every $i \geq 0$, $uv^iwx^iy \in L$.

Proof. Let G be a grammar in Chomsky normal form generating $L \setminus \{\varepsilon\}$, and let k be the number of variables of G. If T is a parsing tree with yield a terminal string w, and the maximum level of a vertex is l, then $|w| \leq 2^{l-1}$. This is easily proved by induction on l. (The right-hand side of a production has length at most 2. If $l = 1$, the minimum possible, the root has a single successor with label $w \in V_T$.)

Put $p = 2^k$. If $z \in L$ and $|z| \geq p$, then $z \in L_G$ and there is an S-tree T (S being the start symbol) with yield z. From the previous paragraph, if l is the maximum level of a vertex of T, then $l \geq k+1$. Let v_0 be the root, and let $v_0, v_1, \ldots, v_l$ be the vertices of a path from v_0 to a vertex of level l. Only v_l can have a terminal as label, so two of the $k+1$ vertices $v_{l-1}, v_{l-2}, \ldots, v_{l-k}, v_{l-k-1}$ must have the same label, say v_r and v_s both have label A, where $l - k - 1 \leq r < s \leq l-1$, so $l - r - 1 \leq k$.

Then v_r is the root of a subtree of T which is an A-tree, say T_1, and v_s is the root of a subtree of T_1 which is also an A-tree, say T_2. Let w be the yield of T_2. Removing T_2 from T_1 (except for v_s) gives an A-tree T_1' with root v_r and yield vAx for some v, x, and the yield of T_1 is vwx. This is illustrated by the following picture.

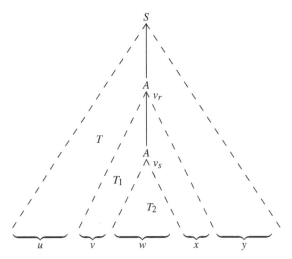

Figure 4.2

Now v_r must have two successors, corresponding to a production $A \longrightarrow BC$, where B, C are variables. For otherwise v_{r+1} would be a leaf, which is impossible as $r+1 \leq s < l$. Both successors are in T_1', hence $|vAx| \geq 2$, so v, x are not both ε.

Similarly, removing T_1 (except for v_r) from T gives an S-tree with yield uAy for some u, y, and the yield of T is $uvwxy = z$. Now v_r has level r, so the maximum level of a vertex of T_1 is $l - r$. Hence $|vwx| \leq 2^{l-r-1} \leq 2^k = p$.

Finally, it follows from Lemma 4.2 that

$$S \xrightarrow[G]{\bullet} uAy, \quad A \xrightarrow[G]{\bullet} vAx, \quad \text{and } A \xrightarrow[G]{\bullet} w.$$

It follows easily by induction on i that $S \xrightarrow[G]{\bullet} uv^iAx^iy$, hence $S \xrightarrow[G]{\bullet} uv^iwx^iy$, for all $i \geq 0$. $\square$

In the proof just given, note that we can obtain an S-tree with yield uv^iwx^iy as follows. Begin with T', the tree obtained by removing T_1 (except for v_r) from T. Add a copy of T_1', identifying its root with v_r. Add another copy of T_1', identifying its root with (the copy of) v_r in the previous copy of T_1'. Repeat, adding a total of i copies of T_1'. (Note that $i = 0$ is allowed, when no copies of T_1' are added and we finish with T'.) Finally, add a copy of T_2, identifying its root with the vertex v_r in the last copy of T_1' (or T', if $i = 0$). If $i \geq 1$, this replaces the final copy of T_1' by a copy of T_1, and if $i = 1$, just results in T. For $i = 2$, the result is illustrated below.

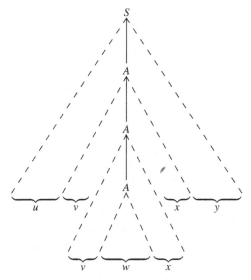

Figure 4.3

We come now to the machines which recognise context-free languages.

Definition. A pushdown stack automaton (abbreviated to PDA) is a septuple

$$M = (Q, F, A, \Gamma, \tau, q_0, z_0)$$

where

(1) Q is a finite set (the set of *states*).
(2) F is a subset of Q (the set of *final* states).
(3) A is a finite set (the *tape alphabet*).
(4) Γ is a finite set (the *stack alphabet*).
(5) τ is a finite subset of $Q \times (A \cup \{\varepsilon\}) \times \Gamma \times Q \times \Gamma^*$ (the set of *transitions*).
(6) $q_0 \in Q$ (the *initial state*).
(7) $z_0 \in Z$ (the *start symbol*).

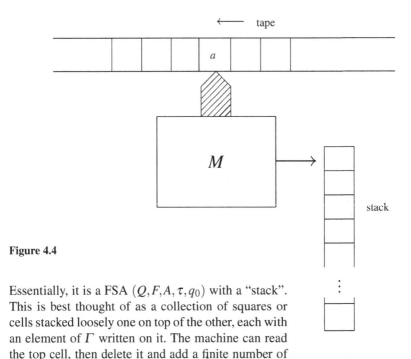

Figure 4.4

Essentially, it is a FSA (Q, F, A, τ, q_0) with a "stack". This is best thought of as a collection of squares or cells stacked loosely one on top of the other, each with an element of Γ written on it. The machine can read the top cell, then delete it and add a finite number of new cells (possibly none) on top of the stack. Exactly what it can do depends on the state, the tape symbol being read, and the stack symbol being read. (The analogy has often been made with the stack of plates sometimes found in cafeterias. These are on top of a spring which ensures that just the top plate is visible. It can either be removed for use, or the person washing dishes can add more plates to the stack.)

Definition. A *configuration* of M is an element of $Q \times A^* \times \Gamma^*$.

The configuration (q, w, γ) is meant to represent the situation that M is in state q, w is the remaining word on the tape at and to the right of the read head, and γ is the word on the stack, read from top to bottom. This is a difference from FSA's, where only the tape symbol being read is needed. We can now formally describe the effect of the transitions.

Definition. If $(q,a,z,q',\alpha) \in \tau$, we say that a configuration $(q,aw,z\beta)$ yields the configuration $(q',w,\alpha\beta)$ by a single move.

Thus if $\alpha = \varepsilon$, the top cell containing z is erased from the stack, otherwise α is added to the top of the stack, replacing z. Note that $a = \varepsilon$ is allowed. This means the machine can operate on the stack, without reading or moving the tape (another difference from a FSA). The following two definitions are just as for Turing machines.

Definition. A *computation* of M, starting at c and ending at c', is a finite sequence of configurations $c = c_1,\ldots,c_n = c'$ (where $n \geq 1$), such that c_i yields c_{i+1} by a single move, for $1 \leq i \leq n-1$.

Definition. If c, c' are configurations, $c \xrightarrow{M} c'$ means there is a computation starting at c, ending at c'.

We can now describe acceptance of words by M. Unlike previous machines, there are two ways this can be done.

Definition. The PDA M accepts $w \in A^*$ by final state if there exists $\gamma \in \Gamma^*$ and $q \in F$ such that $(q_0,w,z_0) \xrightarrow{M} (q,\varepsilon,\gamma)$.

The *language recognised by M by final state*, denoted by $L(M)$, is the set of all elements of A^* accepted by M by final state.

Thus $w \in L(M)$ means that M, started in state q_0, with w on the tape and just z_0 on the stack, has a computation which eventually reaches a final state after reading w on the tape.

Definition. The PDA M accepts $w \in A^*$ by empty stack if there exists $q \in Q$ such that $(q_0,w,z_0) \xrightarrow{M} (q,\varepsilon,\varepsilon)$.

The *language recognised by M by empty stack*, denoted by $N(M)$, is the set of all elements of A^* accepted by M by empty stack.

Thus $w \in N(M)$ if M, started as before, has a computation which eventually results in a configuration with empty stack, after w has been read on the tape.

When considering recognition by empty stack, the set of final states is irrelevant, and is usually taken to be the empty set.

Note. A configuration $c = (q,w,\alpha)$ is called *terminal* if a computation, on reaching c, cannot be continued. That is, there is no transition (q,a,z,q',α) where z is the first letter of α, and a is either ε or the first letter of w. A configuration (q,w,ε) is always terminal (if $\alpha = \varepsilon$, no $z \in \Gamma$ can be the first letter of α). Thus if M empties its stack, it halts.

As with previous machines, there is a notion of deterministic PDA.

Definition. A PDA N is *deterministic* if

(1) For every $q \in Q$, $a \in A \cup \{\varepsilon\}$ and $z \in \Gamma$, there is at most one transition starting with q,a,z.

(2) For every $q \in Q$ and $z \in \Gamma$, if there is a transition starting with q, ε, z, there is no transition starting with q, a, z, for any $a \in A$.

Condition (1) will seem reasonable in view of the definitions for previous machines. Condition (2) prevents a choice between a move without reading the tape and one in which the tape is read.

We now show the equivalence of recognition by final state and recognition by empty stack. If we confine attention to deterministic PDA's these are no longer equivalent.

Theorem 4.11. *If $L = N(M)$ for some PDA M, then $L = L(M')$ for some PDA M'. If M is deterministic, M' can be taken to be deterministic.*

Proof. Suppose $M = (Q, F, A, \Gamma, \tau, q_0, z_0)$. Define $M' = (Q', F', A', \Gamma', \tau', q'_0, x_0)$, where

$$Q' = Q \cup \{q'_0, q'\}, \quad A' = A, \quad \Gamma' = \Gamma \cup \{x_0\}, \quad F' = \{q'\}$$

and τ' consists of all transitions in τ, together with

$$(q'_0, \varepsilon, x_0, q_0, z_0 x_0)$$
$$\text{and } (q, \varepsilon, x_0, q', \varepsilon) \quad \text{for all } q \in Q.$$

Suppose $w \in N(M)$, so $(q_0, w, z_0) \xrightarrow[M]{} (q, \varepsilon, \varepsilon)$ for some $q \in Q$. Hence, using the same sequence of transitions, $(q_0, w, z_0 x_0) \xrightarrow[M]{} (q, \varepsilon, x_0)$. Every transition of M is a transition of M', so there is a computation of M' of the form:

$$(q'_0, w, x_0), (q_0, w, z_0 x_0), \ldots, (q, \varepsilon, x_0), (q', \varepsilon, \varepsilon) \qquad (*)$$

hence $w \in L(M')$. It is easy to see that any computation of M' starting with (q'_0, w, x_0) and ending with $(q', \varepsilon, \gamma)$ for some γ has the form $(*)$. Hence, if $w \in L(M')$, then

$$(q_0, w, z_0 x_0) \xrightarrow[M]{} (q, \varepsilon, x_0)$$

and $(q_0, w, z_0) \xrightarrow[M]{} (q, \varepsilon, \varepsilon)$ by the same transitions, so $w \in N(M)$. If M is deterministic, then clearly M' is.

(The purpose of adding q'_0 is to put x_0 on the bottom of the stack, where it remains while carrying out a computation of M. When M' reads x_0 on the stack, this means M would have emptied the stack without x_0 at the bottom, so M' enters the final state to accept w.)   □

Languages of the form $N(M)$ with M deterministic have a certain property which we now describe. Recall that, if w is a word in some alphabet, and $w = uv$, then u is called a *prefix* of w and v is called a *suffix* of w.

Definition. A language L is *prefix-free* if whenever $w \in L$, no prefix of w, other than w, is in L.

Remark 4.2. If $L = N(M)$ for a deterministic PDA M, then L is prefix-free. For if $w = uv \in L$, u, $v \neq \varepsilon$, $u \in L$, the computation of M accepting w must initially be the same as that accepting u, since M is deterministic. But then M halts after accepting u since its stack is empty, so can't accept w, a contradiction. Note that, if $\varepsilon \in L$, there is a transition starting with q_0, ε, z_0, so no transition starting with q_0, a, z_0 with $a \in A$, by (2) in the definition of deterministic. (As usual, q_0 is the initial state and z_0 the start symbol of M.) It follows that $L = \{\varepsilon\}$, which is prefix-free. Also, if M has one state q_0 and one transition, $(q_0, \varepsilon, z_0, q_0, \varepsilon)$, then M is deterministic and $N(M) = \{\varepsilon\}$.

There are examples of languages of the form $L(M)$, where M is a deterministic PDA, which are not prefix-free, so are not of the form $N(M)$. (See Example 2 near the end of the chapter.) However, the prefix-free property is the only additional requirement needed.

Theorem 4.12. *If $L = L(M)$ for some PDA M, then $L = N(M')$ for some PDA M'. If M is deterministic and L is prefix-free, M' can be taken to be deterministic.*

Proof. Suppose $M = (Q, F, A, \Gamma, \tau, q_0, z_0)$. Define $M' = (Q', F', A', \Gamma', \tau', q_0', x_0)$, where

$$Q' = Q \cup \{q_0', q'\}, \quad A' = A, \quad \Gamma' = \Gamma \cup \{x_0\}, \quad F' = \emptyset$$

and τ' consists of all transitions in τ, together with

$$(q_0', \varepsilon, x_0, q_0, z_0 x_0)$$
$$(q, \varepsilon, z, q', \varepsilon) \quad \text{for all } q \in F, z \in \Gamma'$$
$$(q', \varepsilon, z, q', \varepsilon) \quad \text{for all } z \in \Gamma'.$$

(This time the bottom of stack marker x_0 is needed in case M empties its stack before entering a final state; without it, M' might then accept a word not in $L(M)$. The extra transitions are to make M' empty its stack on entering a final state of M.) The proof that this works is similar to the proof of Theorem 4.11.

Suppose $w \in L(M)$. Then $(q_0, w, z_0) \xrightarrow[M]{} (q, \varepsilon, \gamma)$ for some $q \in F$, $\gamma \in \Gamma^*$, so

$$(q_0, w, z_0 x_0) \xrightarrow[M]{} (q, \varepsilon, \gamma x_0).$$

Since every transition of M is a transition of M', we obtain a computation of M' of the form:

$$(q_0', w, x_0), (q_0, w, z_0 x_0), \ldots, (q, \varepsilon, \gamma x_0), \ldots, (q', \varepsilon, \varepsilon) \qquad (*)$$

hence $w \in N(M')$. Conversely, if $w \in N(M')$, a computation starting with (q_0', w, x_0) and ending with $(p, \varepsilon, \varepsilon)$ for some state p must be of the form $(*)$ (where $q \in F$). For using transitions of M will always leave x_0 on the bottom of the stack, so eventually, after reading w, M' must use a new transition to enter state q', and then it will empty its stack, remaining in state q'. Thus $(q_0, w, z_0 x_0) \xrightarrow[M']{} (q, \varepsilon, \gamma x_0)$ using transitions of M, hence $(q_0, w, z_0) \xrightarrow[M]{} (q, \varepsilon, \gamma)$, so $w \in L(M)$.

Suppose M is deterministic and L is prefix-free. Modify M' by removing all transitions in τ starting with q, for some $q \in F$, to obtain a deterministic PDA M'', whose set of transitions is denoted by τ''. Since $\tau'' \subseteq \tau'$, any computation of M'' is one of M', so $N(M'') \subseteq N(M') = L$. Suppose $w \in L(M) = L$. Then there is some computation

$$(q_0, w, z_0) = c_0, \ldots, c_n = (q, \varepsilon, \alpha) \quad \text{where } q \in F.$$

Let i be the smallest value of j such that $c_j = (q_j, w_j, \alpha_j)$ satisfies $q_j \in F$. Then there is a computation of M'':

$$(q_0', w, z_0), c_0', \ldots, c_i'$$

where c_j' is obtained from c_j by replacing α_j by $\alpha_j x_0$. Further, $w = u w_i$ for some u, and the suffix w_i can be removed from w_j in c_j' to obtain a computation

$$(q_0', u, z_0), c_0'', \ldots, c_i'' \quad \text{of } M''.$$

Since $q_i \in F$, this computation can be continued, without moving the tape, until M'' empties its stack. Hence $u \in N(M'') \subseteq L$. Since L is prefix-free, $u = w \in N(M'')$. □

Theorem 4.13. *If $L = N(M)$ for some PDA M, then L is context-free.*

Proof. Let $M = (Q, F, A, \Gamma, \tau, q_0, z_0)$. Define a grammar $G = (V_N, A, P, S)$ by putting

$$V_N = \{(q, z, p) \mid q, p \in Q, \ z \in \Gamma\} \cup \{S\}$$

and letting P consist of the productions

(1) $S \longrightarrow (q_0, z_0, q)$ for all $q \in Q$
(2) $(q, z, p) \longrightarrow a(q_1, y_1, q_2)(q_2, y_2, q_3) \ldots (q_m, y_m, q_{m+1})$, where $q_{m+1} = p$, for all $q, q_1, \ldots, q_{m+1} \in Q$, all $a \in A \cup \{\varepsilon\}$ and all $z, y_1, \ldots, y_m \in \Gamma$ such that the quintuple $(q, a, z, q_1, y_1 \ldots y_m)$ is a transition. (If $m = 0$, the right-hand side of the production is a.)

The idea is that a leftmost derivation of G, using the productions (2), should simulate a computation of M. Use of a transition of M will lead to use of a corresponding production in a derivation. There are several possible productions, and some means is needed to choose the states q_i which occur in the variables. This needs an interpretation of the variables: (q, z, p) is meant to indicate that, when in state q with z as the top stack symbol, there is a computation ending in state p which "pops" z. This means it has the effect of erasing z from the top of the stack. It does not mean that the final transition used erases z, which may have been replaced earlier by some other string. It means that what is on the stack in state p is what was below z on the stack in state q. (Of course, not all variables will necessarily have this interpretation.) A production (2) is intended to mean that, when M uses the corresponding transition, one way to pop z is to enter state q_1 and pop y_1, ending in state q_2, then pop y_2 ending in state q_3, etc. (Again, not all productions will necessarily have this interpretation.) The terminals occurring at each stage of the derivation will indicate the part of the input that M has read.

We show that this works, by proving that

$$(q,w,z) \xrightarrow[M]{} (p,\varepsilon,\varepsilon) \text{ if and only if } (q,z,p) \xrightarrow[G]{\bullet} w.$$

Suppose that $(q,w,z) \xrightarrow[M]{} (p,\varepsilon,\varepsilon)$. We show by induction on the number of moves of a computation of M that $(q,z,p) \xrightarrow[G]{\bullet} w$. If the number of moves is 1, then w is in $A \cup \{\varepsilon\}$ and (q,w,z,p,ε) is a transition. Therefore $(q,z,p) \longrightarrow w$ is a production, so $(q,z,p) \xrightarrow[G]{\bullet} w$.

Suppose the number of moves, s, is greater than 1. The computation has the form

$$(q,w,z), (q_1,v,y_1 \ldots y_m), \ldots, (p,\varepsilon,\varepsilon)$$

where $w = av$ and $a \in A \cup \{\varepsilon\}$. Let v_1 be the prefix of v such that the stack first becomes as short as $m-1$ symbols after M has read v_1. Let v_2 be the subword of v following v_1 such that the stack first becomes as short as $m-2$ symbols after M has also read v_2, and so on. Thus $v = v_1 \ldots v_m$. Note that, while $v_1 \ldots v_{i-1}$ has been read, $y_i \ldots y_m$ remains on the bottom of the stack.

Let q_i ($i \geq 2$) be the state of M when the stack first becomes as short as $m-i+1$ (so $q_{m+1} = p$). The top stack symbol is then y_i. Thus

$$(q_i,v_i,y_i \ldots y_m) \xrightarrow[M]{} (q_{i+1},\varepsilon,y_{i+1} \ldots y_m)$$

for $1 \leq i \leq m$, by a computation with fewer than s moves. Using the same transitions gives a computation showing $(q_i,v_i,y_i) \xrightarrow[M]{} (q_{i+1},\varepsilon,\varepsilon)$. It follows by induction that $(q_i,y_i,q_{i+1}) \xrightarrow[G]{\bullet} v_i$ for $1 \leq i \leq m$. From the first move in the computation, there is a production

$$(q,z,p) \longrightarrow a(q_1,y_1,y_2)(q_2,y_2,q_3) \ldots (q_m,y_m,q_{m+1}).$$

Hence, there is a G-derivation:

$$(q,z,p), a(q_1,y_1,y_2)(q_2,y_2,q_3) \ldots (q_m,y_m,q_{m+1}), \ldots,$$
$$av_1(q_2,y_2,q_3) \ldots (q_m,y_m,q_{m+1}), \ldots,$$
$$av_1v_2(q_3,y_3,q_4) \ldots (q_m,y_m,q_{m+1}), \ldots, av_1v_2 \ldots v_m = w$$

as required. (Note that, if we take leftmost derivations of v_i from (q_i,y_i,q_{i+1}), the result is a leftmost derivation of w.)

Conversely, assume $(q,z,p) \xrightarrow[G]{\bullet} w$. We show that $(q,w,z) \xrightarrow[M]{} (p,\varepsilon,\varepsilon)$ by induction on the number of steps in a derivation of w from (q,z,p). If this number is 1, then $(q,z,p) \longrightarrow w$ is a production. This can only happen if $w \in A \cup \{\varepsilon\}$ and there is a transition (q,w,z,p,ε), hence $(q,w,z) \xrightarrow[M]{} (p,\varepsilon,\varepsilon)$.

Suppose the number of steps in the derivation is greater than 1. The derivation
has the form

$$(q,z,p), a(q_1,y_1,q_2)(q_2,y_2,q_3)\dots(q_m,y_m,q_{m+1}),\dots,w$$

where $q_{m+1} = p$ and $(q,a,z,q_1,y_1\dots y'_m)$ is a transition. We can write $w = av_1\dots v_m$
where, for $1 \le i \le m$, $(q_i,y_i,q_{i+1}) \xrightarrow[G]{\bullet} v_i$, and by induction

$$(q_i,v_i,y_i) \xrightarrow[M]{} (q_{i+1},\varepsilon,\varepsilon)$$

for all such i. Thus, for each i, using exactly the same transitions, we find that

$$(q_i,v_iv_{i+1}\dots v_m,y_iy_{i+1}\dots y_m) \xrightarrow[M]{} (q_{i+1},v_{i+1}\dots v_m,y_{i+1}\dots y_m).$$

Hence there is a computation

$$(q,w,z),(q_1,v_1v_2\dots v_m,y_1y_2\dots y_m),\dots,(q_2,v_2\dots v_m,y_2\dots y_m),\dots,$$
$$(q_3,v_3\dots v_m,y_3\dots y_m),\dots,(q_{m+1},\varepsilon,\varepsilon)$$

and since $q_{m+1} = p$, this completes the induction.

Finally, it follows that $(q_0,z_0,p) \xrightarrow{\bullet} w$ if and only if $(q_0,w,z_0) \xrightarrow[M]{} (p,\varepsilon,\varepsilon)$. Now
using the productions (1), it follows that $S \xrightarrow{\bullet} w$ if and only if $(q_0,w,z_0) \xrightarrow[M]{} (p,\varepsilon,\varepsilon)$
for some $p \in Q$. hence $L_G = N(M)$.   □

Remark 4.3. With more care, the proof shows that leftmost derivations of G sim-
ulate computations of M in a precise manner. Given a computation $(q_0,w,z_0) = c_1,c_2,\dots,c_n = (p,\varepsilon,\varepsilon)$, there is a unique associated leftmost derivation $(q_0,z_0,p) = \alpha_1,\alpha_2,\dots,\alpha_n = w$, and any leftmost derivation starting with (q_0,z_0,p) arises in this
way. If $c_i = (q_{i-1},v_i,\beta_i)$ and $\beta_i = z_{i1}\dots z_{ik_i}$, where $z_{ij} \in Z$, then

$$\alpha_i = u_i(-,z_{i1},-)\dots(-,z_{ik_i},-)$$

where $w = u_iv_i$, and the dashes represent certain elements of Q.

Suppose M is deterministic; it follows that G is unambiguous. Further, in configuration c_i, u_i has been read from the tape. When u_i is first read, the computation
next uses a uniquely determined sequence of transitions (possibly none) of the form
$(q,\varepsilon,-,-,-)$ before reading the next symbol on the tape. This sequence will appear in any computation in which u_i is read from the tape at some point. Thus if
$(q_0,z_0,p) = \alpha_1,\alpha_2,\dots,\alpha_n = w$ is a leftmost derivation with $\alpha_j = u_i\gamma$ for some j (γ
is a string of variables of G), the corresponding computation of M will use all of this
sequence of transitions, and all of the corresponding words (beginning with u_i) will
appear in the derivation. Hence, given two leftmost derivations $S,(q_0,z_0,p),\dots,w$
and $S,(q_0,z_0,p),\dots,w'$, if $u\gamma$ appears in one derivation and $u\gamma'$ appears in the other,
then both words appear in both derivations. (Here $u \in V_T^*$, $\gamma, \gamma' \in V_N^*$.)

Theorem 4.14. *If L is context-free, then $L = N(M)$ for some PDA M.*

Proof. Suppose first that $\varepsilon \notin L$. Let $L = L_G$ where $G = (V_N, V_T, P, S,)$ is a context-free grammar in Greibach normal form. Let $M = (\{q\}, \emptyset, V_T, V_N, \tau, q, S)$, where τ consists of all (q, a, A, q, γ) for all productions $A \longrightarrow a\gamma$ in P. For $\alpha \in V_N^*$ and $w \in V_T^*$, we show that

$$S \xrightarrow{\bullet} w\alpha \text{ if and only if } (q, w, S) \xrightarrow[M]{} (q, \varepsilon, \alpha).$$

Suppose $S \xrightarrow{\bullet} w\alpha$, so there is a leftmost derivation of w from S. We show by induction on the number of steps in this derivation that $(q, w, S) \xrightarrow[M]{} (q, \varepsilon, \alpha)$. If the number of steps is 0, then $w = \varepsilon$, $\alpha = S$, and $(q, \varepsilon, S) \xrightarrow[M]{} (q, \varepsilon, S)$ by 0 moves. Otherwise, the derivation has the form

$$S, \ldots, vA\beta, va\gamma\beta$$

where $v \in V_T^*$, $\beta \in V_N^*$ and $A \longrightarrow a\gamma$ is a production. Thus $w = va$ and $\alpha = \gamma\beta$. By induction, $(q, v, S) \xrightarrow[M]{} (q, \varepsilon, A\beta)$, so $(q, w, S) \xrightarrow[M]{} (q, a, A\beta)$. Also, (q, a, A, q, γ) is a transition. Hence there is a computation

$$(q, w, S), \ldots, (q, a, A\beta), (q, \varepsilon, \gamma\beta)$$

as required.

Conversely, suppose $(q, w, S) \xrightarrow[M]{} (q, \varepsilon, \alpha)$. we show by induction on the number of moves in a corresponding computation that $S \xrightarrow{\bullet} w\alpha$. This is obvious if the number of moves is 0. Otherwise, put $w = va$; the computation has the form

$$(q, va, S), \ldots, (q, a, \beta'), (q, \varepsilon, \alpha).$$

The final transition used comes from a production of the form $A \longrightarrow a\gamma$, so $\beta' = A\beta$ for some β, and $\alpha = \gamma\beta$. Using all but the final transition, we obtain a computation

$$(q, v, S), \ldots (q, \varepsilon, \beta')$$

so by induction $S \xrightarrow{\bullet} v\beta' = vA\beta$. Also, $vA\beta \xrightarrow{\bullet} va\gamma\beta = w\alpha$, hence $S \xrightarrow{\bullet} w\alpha$.

Taking $\alpha = \varepsilon$ gives $S \xrightarrow{\bullet} w$ if and only if $(q, w, S) \xrightarrow[M]{} (q, \varepsilon, \varepsilon)$, hence $L = N(M)$.

Finally, if $\varepsilon \in L$, then $L \setminus \{\varepsilon\}$ is context-free (by Cor. 1.2), so by what we have proved, $L \setminus \{\varepsilon\} = N(M)$ for some PDA M with initial state q and start symbol S. Add a new state q' to M, and a new transition $(q, \varepsilon, S, q', \varepsilon)$ to obtain a PDA M' with $L = N(M')$. $\qquad \square$

We now have two new classes of languages: those which are $L(M)$ for some deterministic PDA M, and those which are $N(M)$ for some deterministic PDA M. We shall show that these can be defined by corresponding classes of grammars. We begin by giving a name to these classes

Definition. A language L is *deterministic* if $L = L(M)$ for some deterministic PDA M, and L is *strict deterministic* if $L = N(M)$ for some deterministic PDA M.

Before proceeding, we prove two results concerning regular, context-free and deterministic languages.

Lemma 4.15. *A regular language is deterministic.*

Proof. Let $M = (Q, F, A, \tau, q_0)$ be a deterministic FSA recognising the regular language L. Let M' be the deterministic PDA $(Q, F, A, \{z_0\}, \tau', q_0, z_0)$, where τ' consists of all transitions (q, a, z_0, q', z_0) for $(q, a, q') \in \tau$. It is easily shown that if $w = a_1 \ldots a_n \in A^*$ ($n \geq 0$), then there is a computation of M with label w ending at state q if and only if $(q_0, w, z_0) \xrightarrow[M']{} (q, \varepsilon, z_0)$. (The proof is by induction on $|w|$.) Since M and M' have the same set of final states, $L = L(M')$.   □

Recall from Exercise 6, Chapter 1 that the intersection of two context-free languages is not necessarily context-free. However, we can now prove the following.

Lemma 4.16. *Let R be a regular language. If L is a context-free language, then $L \cap R$ is context-free. If L is deterministic, then $L \cap R$ is deterministic.*

Proof. We can assume L, R have the same alphabet, A (otherwise take the union of their alphabets as the new alphabet). By Theorem 4.14 and Theorem 4.11, $L = L(M)$ for some PDA M, say $M = (Q, F, A, \Gamma, \tau, q_0, z_0)$. Also, R is recognised by some deterministic FSA, say $M' = (Q', F', A, \tau', q_0')$.

Let δ be the transition function of M'. Define a new PDA M'' by

$$M'' = (Q \times Q', F \times F', A, \Gamma, \tau'', (q_0, q_0'), z_0)$$

where, for each $(q, a, z, p, \alpha) \in \tau$ and $q' \in Q'$, τ'' contains the transition

$$((q, q'), a, z, (p, \delta(q', a)), \alpha).$$

(Recall that $a \in A \cup \{\varepsilon\}$, and $\delta(q', \varepsilon) = q'$.) It is left to the reader to verify that $L \cap R = L(M'')$. If M is deterministic, then clearly M'' is, and the last part of the lemma follows.   □

We now define the classes of grammars which will be used to characterise the two language classes recognised by deterministic PDA's. In what follows, we make some notation conventions. Greek letters denote elements of $(V_N \cup V_T)^*$, lower case letters denote elements of $(V_T \cup \{\$\})^*$, where $\$$ is a new letter not in $V_N \cup V_T$, and upper case letters denote elements of V_N.

Definition. Let $k \in \mathbb{N}$ and let $G = (V_N, V_T, P, S)$ be a context-free grammar. Let $\$$ be a letter not in $V_N \cup V_T$. Then G is called $LR(k)$ if S does not appear on the right-hand side of any production, and given rightmost P-derivations

$$S\$^k, \ldots, \alpha A w_1 w_2, \alpha \beta w_1 w_2$$

$$S\$^k, \ldots, \gamma B w, \alpha \beta w_1 w_3$$

where $|w_1| = k$, then $\gamma = \alpha$, $A = B$, and $w = w_1 w_3$.

(In a rightmost derivation of two words from $S\$^k$ which agree up to k letters beyond the point of the last replacement, the words at the penultimate step agree up to k symbols beyond the point of the last replacement. The new letter $\$$ is used as a "padding symbol", to make sure there are k letters beyond the point of the last replacement. Note that, in these derivations, $\$^k$ always remains at the right-hand end, as $\$$ does not occur in any production. In particular, $w_1 w_2$ and $w_1 w_3$ end in $\$^k$. The term $LR(k)$ stands for something like "parsing from the left of rightmost derivations with k steps of lookahead".)

Let $G = (V_N, V_T, P, S)$ be a context-free grammar with r productions, and number the productions of G from 1 to r. For $k \geq 0$, $w \in V_T^* \{\$\}^*$ with $|w| = k$ and $1 \leq i \leq r$, let $R_k(i, w)$ be the set of words γ for which there is a rightmost derivation

$$S\$^k, \ldots, \alpha B w w_2, \alpha \beta w w_2$$

where $B \longrightarrow \beta$ is the ith production and $\gamma = \alpha \beta w$.

Lemma 4.17. *Let $G = (V_N, V_T, P, S)$ be a context-free grammar. Then for any $k \geq 0$, $w \in V_T^* \{\$\}^*$ with $|w| = k$ and $1 \leq i \leq r$, the set $R_k(i, w)$ is regular.*

Proof. Define a grammar $G' = (V_N', V_N \cup V_T \cup \{\$\}, P', S')$ as follows. The elements of V_N' are the ordered pairs (A, v), where $A \in V_N$, $v \in V_T^* \{\$\}^*$ and $|v| = k$. The start symbol S' is $(S, \$^k)$. The productions in P' are as follows.

(1) Suppose $A \longrightarrow X_1 \ldots X_n$ is in P (here $X_i \in V_N \cup V_T$). If $1 \leq j \leq n$ and $X_j \in V_N$, then P' contains the productions

$$(A, v) \longrightarrow X_1 \ldots X_{j-1} (X_j, v')$$

for every v, v' in $V_T^* \{\$\}^*$ of length k such that for some v'', $X_{j+1} \ldots X_n v \xrightarrow{\bullet}_{G} v'v''$.

(2) If $B \longrightarrow \beta$ is the ith production, then the production $(B, w) \longrightarrow \beta w$ is in P'.

We show that $L_{G'} = R_k(i, w)$. (This will not finish the proof-G' has to be modified to obtain a regular grammar.) In a G'-derivation, all strings occurring are of the form $\alpha(A, v)$, where $\alpha \in (V_N \cup V_T)^*$, except possibly the last one; the production in (2) can be used only once, as the final step in the derivation. Thus, it suffices to show that, for $v \in V_T^* \{\$\}^*$ of length k,

$(S, \$^k) \xrightarrow[G']{\bullet} \alpha(A, v)$ if and only if, for some v_1, there is a rightmost P-derivation

$$S\$^k, \ldots, \alpha A v v_1.$$

Assume there is a rightmost P-derivation $S\$^k, \ldots, \alpha A v v_1$. We show by induction on the number of steps in the derivation that $(S, \$^k) \xrightarrow[G']{\bullet} \alpha(A, v)$. If the number of steps is 1, this is easy to see. If the number of steps is greater than 1, then it has the form

$$S\$^k, \ldots, \gamma C v' v_2, \gamma \delta v' v_2 = \alpha A v v_1$$

where $v' \in V_T^*\{\$\}^*$ has length k and the final production used is $C \rightarrow \delta$. Suppose $\delta \notin V_T^*$. Then δ has the form $\delta' A v_3$, where $\gamma \delta' = \alpha$ and $v_3 v' v_2 = v v_1$. It follows that v is a prefix of $v_3 v'$, so $(C, v') \longrightarrow \delta'(A, v)$ is a production of P'. By induction, $(S, \$^k) \xrightarrow[G']{\bullet} \gamma(C, v')$. Hence $(S, \$^k) \xrightarrow[G']{\bullet} \gamma \delta'(A, v) = \alpha(A, v)$.

Otherwise ($\delta \in V_T^*$), γ has the form $\alpha A y$, where $y \in V_T^*$. At some point in the derivation, the letter A was introduced, by a production of the form $D \longrightarrow \gamma_1 A \gamma_2$, so it has the form

$$S\$^k, \ldots, \gamma_3 D y_1 y', \gamma_3 \gamma_1 A \gamma_2 y_1 y', \ldots, \alpha A v v_1$$

where $y_1 \in V_T^*\{\$\}^*$ and $|y_1| = k$. Thus $\alpha = \gamma_3 \gamma_1$ since the derivation is rightmost, and inductively $(S, \$^k) \xrightarrow[G']{\bullet} \gamma_3(D, y_1)$. It follows that $\gamma_2 \xrightarrow[P]{\bullet} y_2$ for some terminal string y_2 such that $y_2 y_1$ is a prefix of $v v_1$. Hence $(D, y_1) \longrightarrow \gamma_1(A, v)$ is in P'. Thus $(S, \$^k) \xrightarrow[G']{\bullet} \gamma_3 \gamma_1(A, v) = \alpha(A, v)$ as required.

To prove the converse, it suffices by Lemma 4.2 to show that if $(S, \$^k) \xrightarrow[G']{\bullet} \alpha(A, v)$ then for some v_1, there is a P-derivation $S\$^k, \ldots, \alpha A v v_1$. The proof is by induction on the number of steps in a derivation of $\alpha(A, v)$ from $(S, \$^k)$, and is left to the reader.

Finally, we have to convert G' to a regular grammar generating $R_k(i, w)$. All productions of G' are of the form $A \longrightarrow vB$ or $A \longrightarrow v$ where v is a string of terminals of G'. Applying the procedure of Lemmas 4.1 and 4.6 gives a grammar of the same form generating $R_k(i, w)$, where all strings v which occur have length at least 1, except that $S' \longrightarrow \varepsilon$ may be present. If $A \rightarrow v$ is a production, where $v = x_1 \ldots x_n$ ($n \geq 2$), add new variables $B_1, \ldots, B_{n-1}$ and replace this production by the productions

$$A \longrightarrow x_1 B_1, B_1 \longrightarrow x_2 B_2, \ldots, B_{n-2} \longrightarrow x_{n-1} B_{n-1}, B_{n-1} \longrightarrow x_n.$$

The variables $B_1, \ldots B_{n-1}$ can be added one by one. First add B_1 and replace $A \longrightarrow v$ by $A \longrightarrow x_1 B_1$ and $B_1 \longrightarrow x_2 \ldots x_n$, then (if $n > 2$) add B_2 and replace $B_1 \longrightarrow x_2 \ldots x_n$ by $B_1 \longrightarrow x_2 B_2$ and $B_2 \longrightarrow x_3 \ldots x_n$ and so on (the final step being to add B_{n-1} and replace $B_{n-2} \longrightarrow x_{n-1} x_n$ by $B_{n-2} \longrightarrow x_{n-1} B_{n-1}$ and $B_{n-1} \longrightarrow x_n$. It follows that the new grammar generates $R_k(i, w)$ (see Exercise 3 at the end of the chapter).

If a production has the form $A \longrightarrow vB$ ($v = x_1 \ldots x_n$, $n \geq 2$) proceed similarly, but the last production should be $B_{n-1} \longrightarrow x_n B$. Again the variables can be added one by one (changing the production $B_1 \longrightarrow x_2 \ldots x_n$ to $B_1 \longrightarrow x_2 \ldots x_n B$, etc) and by Exercise 3, the new grammar generates $R_k(i, w)$.

Doing this for every such production gives a regular grammar which generates $R_k(i, w)$.   □

Remark 4.4. A grammar is called *right linear* if all productions are of the form $A \longrightarrow uB$ or $A \longrightarrow u$, where A, B are variables and u is a string of terminals. The last part of the preceding proof shows that a right linear grammar generates a regular language. Similarly, one can define left linear (all productions of the form $A \longrightarrow Bu$ or $A \longrightarrow u$), and show that a left linear grammar generates the same language as some left regular grammar (see Remark 1.1). In view of Remark 1.1, the following are equivalent, for a language L.

(1) L is regular.
(2) L is generated by a right linear grammar.
(3) L is generated by a left linear grammar.

Remark 4.5. Let $k \geq 0$. In the circumstances of Lemma 4.17, suppose G is $LR(k)$ and $\gamma \in R_k(i,w)$ and $\gamma u \in R_k(j,v)$. Then $i = j$, $v = w$ and $u = \varepsilon$. (Recall the notation convention: $u \in (V_T \cup \{\$\})^*$.) For there are derivations

$$S\$^k, \ldots, \alpha A w w_2, \alpha \beta w w_2 \quad \text{and} \quad S\$^k, \ldots, \delta B v v_2, \delta \zeta v v_2,$$

where $\gamma = \alpha \beta w$ and $\gamma u = \delta \zeta v$, and $A \longrightarrow \beta, B \longrightarrow \zeta$ are respectively the ith and jth productions of G. By the $LR(k)$ assumption, $\alpha = \delta$, $A = B$ and $wuv_2 = vv_2$, hence $w = v$ as $|w| = |v| = k$. It follows that $u = \varepsilon$ and $\beta = \zeta$, so $i = j$.

Now suppose G is $LR(k)$. By Theorem 1.4, there is a FSA $M_k(i,w)$ with alphabet $V_N \cup V_T \cup \{\$\}$, recognising $R_k(i,w)$, for every possible value of i and w. We can assume that for $(i,w) \neq (j,v)$, $M_k(i,w)$ and $M_k(j,v)$ have no states in common. Let $R_k = \bigcup_{i,w} R_k(i,w)$. From the proof of Lemma 1.5(3), there is a FSA M_k with alphabet $V_N \cup V_T \cup \{\$\}$ recognising R_k. Its transition diagram is constructed by taking the union of the transition diagrams of $M_k(i,w)$ for each value of i and w, then adding a new state s as initial state, with extra edges from s labelled ε to the initial state of $M_k(i,w)$, for each i and w. The final states are those of every $M_k(i,w)$.

Now apply the construction of Prop.1.3 to M_k, to obtain a deterministic FSA D_k recognising R_k. The states of D_k are subsets of the states of M_k, and a state is a final state if and only if it contains a final state of M_k. Suppose there is a path in the transition diagram of D_k from the initial state to a final state Q, with label γ. Assume Q contains a final state q of $M_k(i,w)$ and a final state q' of $M_k(j,v)$. It is easily seen that there are paths in $M_k(i,w)$ and $M_k(j,v)$ starting at their initial states and ending at q, q' respectively, both with label γ. Thus $\gamma \in R_k(i,w)$ and $\gamma \in R_k(j,v)$, so by Remark 4.5, $i = j$.

Next, modify D_k, to obtain D'_k, by letting the final states of D'_k be the final states Q of D_k for which there is a path in the transition diagram of D_k from the initial state to Q. Then D'_k is still deterministic and recognises R_k, and we have associated to each final state of D'_k a unique production of G.

Theorem 4.18. *If $L = L_G$ for an $LR(k)$ grammar G, then $L\$^k$ is deterministic (where $\$$ is a letter not in the alphabet of L).*

Proof. Number the productions. Let R_k be the set in the preceding discussion, and let D be the deterministic FSA D'_k recognising R_k, with alphabet $V_N \cup V_T \cup \{\$\}$. Denote the initial state of D by d. By an edge of D, we mean an edge of its transition diagram. Let $\lambda(e)$ denote the label on edge e of D. If $u = e_1, \ldots e_n$ is a sequence of edges of D (not necessarily a path), define the label on u, $\lambda(u)$, to be $\lambda(e_1) \ldots \lambda(e_n)$

and put $t(u) = t(e_n)$ $(t(\varepsilon) = d)$. To construct a deterministic PDA M recognising $L\$^k$, we take as tape alphabet $(V_T \cup \{\$\})$, and as stack alphabet we take the set of edges of D together with a start symbol z_0. Note that z_0 will be used as a bottom of stack marker. We define $t(z_0)$ to be d. There is an initial state q_0. The machine carries out one of the following two steps as often as possible.

(1) In state q_0, if the top symbol x of the stack is such that $t(x)$ is a final state of D, read symbols from the stack, storing them in the states as a word, with the top symbol on the right. At most $k+l$ symbols are read, where l is the length of the longest right-hand side of a production. Suppose, during the computation, the label on the word read is of the form βw, where $|w| = k$ and the production associated to $t(e)$ is of the form $A \longrightarrow \beta$. Let the symbol on top of the stack after reading βw be z. Add new edges $f_0, f_1, \ldots, f_k$ on top of the stack (f_0 at the bottom), where f_0 is the edge from $t(z)$ with label A, and for $i > 0$, f_i is the edge from $t(f_{i-1})$ with label a_i, where $w = a_1 \ldots a_k$. Then return to state q_0 without further altering the stack or reading the tape. If this never happens, the machine will halt after reading at most $k+l$ symbols.

   To do this, take a new letter q_1, and add new states (q_1, u), where u is a word of length at most $k+l$ in the edges of D. Add transitions $(q_0, \varepsilon, x, (q_1, \varepsilon), x)$ for x in the stack alphabet and $t(x)$ a final state of D, and $((q_1, u), \varepsilon, e, (q_1, eu), \varepsilon)$, for u of length less than $k+l$, where e is an edge of D and one of the following fails.

   (a) $\lambda(u)$ is of the form βw where $|w| = k$.
   (b) $t(u)$ is a final state of D.
   (c) the production associated to $t(u)$ has the form $A \longrightarrow \beta$.

   For every u satisfying (a)–(c) of length at most $k+l$, add a transition

   $$((q_1, u), \varepsilon, z, q_0, f_k \ldots f_1 f_0 z)$$

   where z is in the stack alphabet, f_0 is the edge of D from $t(z)$ with label A and for $i > 0$, f_i is the edge from $t(f_{i-1})$ with label a_i, where $w = a_1 \ldots a_k$.

(2) In state q_0, if the top symbol x of the stack is such that $t(x)$ is not a final state of D, read the first $k+1$ symbols of the stack (or as many as possible if there are fewer than $k+1$ symbols on the stack), storing them in the states as a word (with the top symbol on the right). If the word obtained is u, where $\lambda(u) = S\$^k$, and the top symbol of the stack is z_0, move to a final state (only one final state is needed). Otherwise, restore the stack. If possible, read a symbol from the tape, say a, and add f to the top of the stack, where f is the edge from $t(x)$ with label a. Then return to state q_0.

   To do this, add a new symbol q_2 and states (q_2, u) where u is a word of length at most $k+1$ whose letters are edges of D. Add transitions $(q_0, \varepsilon, z, (q_2, \varepsilon), z)$ for z in the stack alphabet and $t(z)$ not a final state of D, and $((q_2, u), \varepsilon, e, (q_2, eu), \varepsilon)$, for u of length less than $k+1$ and e an edge of D. Also, add a state p as the only final state and transitions $((q_2, u), \varepsilon, z_0, p, z_0)$, whenever $\lambda(u) = S\$^k$.

Now add new states (q_3, u) where u is a word of length at most $k + 1$ whose letters are edges of D. Add transitions

$$((q_2, u), \varepsilon, z, (q_3, u), z)$$

for z in the stack alphabet, and u of length $k + 1$, except when $\lambda(u) = S\$^k$ and $z = z_0$. Also add transitions $((q_2, u), \varepsilon, z_0, (q_3, u), z_0)$ for u with $|u| < k + 1$. Then add transitions

$$((q_3, eu), \varepsilon, z, (q_3, u), ez)$$

where e is an edge of D, $|u| \leq k$ and z in the stack alphabet. Finally add transitions $((q_3, \varepsilon), a, x, q_0, fx)$, where a is in the tape alphabet, x is in the stack alphabet and f is the edge from $t(x)$ with label a.

It is left to the reader to check that M is deterministic. Suppose M is started in state q_0 with w on the tape and z_0 on the stack. Whenever M is in state q_0, the stack contains $z_0 e_1 \ldots e_n$, (with z_0 at the bottom), where $e_1, \ldots, e_n$ are the edges in a path starting at d. This follows by induction on the number of moves. Call $\lambda(e_1) \ldots \lambda(e_n)$ the *label* on the stack. If, in state q_0, the label γ on the stack is in R_k, then by Remark 4.5, $\gamma \in R_k(i, w)$ for unique i and w. By the discussion preceding the theorem, the ith production is the production associated to $t(x)$, where x is the top symbol of the stack. If this ith production is $A \longrightarrow \beta$, then γ has the form $\alpha\beta w$. Step 1 is carried out and the stack label becomes $\alpha A w$. Otherwise, either M halts during Step 1, or Step 2 is carried out. Then either M enters the final state, or an edge corresponding to the tape symbol being read is added to the top of the stack (or if there is no symbol on the tape, the machine halts in state (q_3, ε)).

If $w \in L\$^k$, there is a rightmost derivation of w from $S\$^k$. During the computation, when Step 1 is carried out for the ith time, let the stack label initially be α_i (so $\alpha_i \in R_k$) and let u_i be the remaining word on the tape. Suppose Step 1 is carried out r times. Then this derivation is $S\$^k, \alpha_r u_r, \alpha_{r-1} u_{r-1}, \ldots, \alpha_1 u_1$, and the productions used are those used in Step 1, in reverse order. This follows by induction on r, using Remark 4.5, and is left to the reader. If the first production used is $S \longrightarrow \beta$, then $\alpha_r u_r = \beta\$^k$, and $\beta\$^k \in R_k$. Again by Remark 4.5, $\alpha_r = \beta\$^k$ and $u_r = \varepsilon$. Thus after the final use of Step 1, the stack label is $S\$^k$ and all of w has been read. When the stack label is $S\$^k$, Step 2 is carried out and M enters the final state. This is because $S\$^k \notin R_k$, as S does not occur on the right of any production. Thus M accepts w. Conversely, if M accepts, then $S\$^k$ followed by the words $\alpha_i u_i$ as defined above, in reverse order, give a derivation of w from $S\$^k$, hence $w \in L\$^k$. It follows that M recognises $L\$^k$. $\square$

Theorem 4.19. *If L is strict deterministic, then $L = L_G$ for some LR(0) grammar G.*

Proof. Let M be a deterministic PDA with $L = N(M)$. Construct the grammar obtained from M in Theorem 4.13, then use Lemma 4.3 to remove variables and productions so that all variables are generating, obtaining a grammar G. We show that G is LR(0). It is clear that S does not occur on the right-hand side of any production of G. Suppose there are rightmost derivations

$$S, \ldots, \alpha A w_2, \alpha \beta w_2 \tag{4.1}$$

$$S, \ldots, \gamma B w, \gamma \delta w = \alpha \beta w_3 \tag{4.2}$$

as in the definition of $LR(0)$. We have to show that $\gamma = \alpha$, $A = B$, and $w = w_3$. By symmetry we can assume that $|\gamma \delta| \leq |\alpha \beta|$, so we can write $\gamma \delta u = \alpha \beta$ and $w = u w_3$ for some u. Let w_α, w_β be terminal strings with $\alpha \overset{\bullet}{\longrightarrow} w_\alpha$, $\beta \overset{\bullet}{\longrightarrow} w_\beta$. By Lemma 4.2, there are rightmost derivations of w_α, w_β from α, β respectively, hence the derivation 4.1 can be extended to a rightmost derivation

$$S, \ldots, \alpha A w_2, \alpha \beta w_2, \ldots, \alpha w_\beta w_2, \ldots, w_\alpha w_\beta w_2 \tag{4.3}$$

and this derivation comes from a parsing tree, which determines a corresponding leftmost derivation, which has the form

$$S, \ldots, w_\alpha A \zeta_2, w_\alpha \beta \zeta_2, \ldots, w_\alpha w_\beta \zeta_2, \ldots, w_\alpha w_\beta w_2 \tag{4.4}$$

where ζ_2 is a string with $\zeta_2 \overset{\bullet}{\longrightarrow} w_2$. It is easy to see that there are terminal strings w_γ, w_δ, such that $\gamma \overset{\bullet}{\longrightarrow} w_\gamma$, $\delta \overset{\bullet}{\longrightarrow} w_\delta$ and $w_\gamma w_\delta u = w_\alpha w_\beta$. Again by Lemma 4.2, derivation 4.2 can be extended to a rightmost derivation of the form:

$$S, \ldots, \gamma B u w_3, \gamma \delta u w_3, \ldots, w_\gamma w_\delta u w_3 \tag{4.5}$$

and this derivation corresponds to a parsing tree which determines a leftmost derivation of the form:

$$S, \ldots, w_\gamma w_\delta u \zeta_3, \ldots, w_\gamma w_\delta u w_3 \tag{4.6}$$

where ζ_3 is a string with $\zeta_3 \overset{\bullet}{\longrightarrow} w_3$. By Remark 4.3, the string $w_\gamma w_\delta u \zeta_3$ occurs in the derivation 4.4 and so $\zeta_3 \overset{\bullet}{\longrightarrow} w_2$. The first production in derivation 4.6 has the form $S \longrightarrow \varphi \psi$, where $\varphi \overset{\bullet}{\longrightarrow} w_\gamma w_\delta u$ and $\psi \overset{\bullet}{\longrightarrow} \zeta_3$. The rightmost derivation 4.5 has the form

$$S, \varphi \psi, \ldots, \varphi w_3, \ldots, \gamma B u w_3, \gamma \delta u w_3, \ldots, w_\gamma w_\delta u w_3$$

and so $\varphi \overset{\bullet}{\longrightarrow} \gamma B u$. By Lemma 4.2, there is a rightmost derivation of the form

$$S, \varphi \psi, \varphi \zeta_3, \ldots, \varphi w_2, \ldots, \gamma B u w_2, \gamma \delta u w_2, \ldots, w_\gamma w_\delta u w_2 \tag{4.7}$$

Truncating this derivation gives a rightmost derivation

$$S, \ldots, \gamma B u w_2, \gamma \delta u w_2 = \alpha \beta w_2 \tag{4.8}$$

By Remark 4.3, G is unambiguous, so derivations 4.3 and 4.7 are equal, hence derivations 4.1 and 4.8 are the same. (The string $\alpha \beta w_2$ cannot occur twice in derivation 4.3, otherwise we could obtain a shorter rightmost derivation of $w_\alpha w_\beta w_2$ from S, contradicting the fact that G is unambiguous.) In particular, $\alpha A w_2 = \gamma B u w_2$, so $u w_2 = w_2$, $A = B$ and $\alpha = \gamma$, hence $u = \varepsilon$ and $w = w_3$. $\qquad \square$

Lemma 4.20. *Let* $G = (V_N, V_T, P, S)$ *be a context-free grammar.*

(1) *If* G *is* $LR(k)$ *for some* k *then* G *is unambiguous.*
(2) *If* G *is* $LR(0)$ *then* L_G *is prefix-free.*

Proof. (1) Given a rightmost derivation $S, \ldots, \beta, \alpha$ $(\alpha \in (V_N \cup V_T)^*)$, let \$ be a letter not in V_T. Adding $\$^k$ to the right of every string in the derivation gives a rightmost derivation. Taking $w_2 = w_3$ in the definition of $LR(k)$ we find that $\beta\$^k$ is uniquely determined by $\alpha\$^k$, hence β is uniquely determined by α. Also, there is no derivation $S, \ldots, S$ of length greater than 1, as S does not appear on the right of any production. Hence a rightmost derivation of α from S is uniquely determined by α, by induction on its length.

(2) Suppose $u, uv \in L_G$ and $v \neq \varepsilon$. There are rightmost derivations

$$S, \ldots, \alpha Aw, \alpha\beta w = u$$
$$S, \ldots, \gamma Bw', \alpha\beta wv.$$

By the $LR(0)$ condition, $\gamma Bw' = \alpha Awv$. Removing the last words in the derivations gives rightmost derivations and this argument can be repeated. Continuing, we find that Sv eventually appears in the second derivation, so $S \xrightarrow{\bullet} Sv$. But this is impossible as $v \neq \varepsilon$ and S does not appear on the right-hand side of any production. $\qquad\square$

Theorem 4.21. *For a language* L, *the following are equivalent.*

(1) $L = L_G$ *for some* $LR(0)$ *grammar* G.
(2) L *is deterministic and prefix-free.*
(3) L *is strict deterministic.*

Proof. Assume (1). By Theorem 4.18, L is deterministic and by Lemma 4.20, it is prefix-free, so (2) holds. It follows from Theorem 4.12 that (2) implies (3), and from Theorem 4.19 that (3) implies (1). $\qquad\square$

To deal with $LR(k)$ languages in general, some digressions are required. First, we introduce a new operation on languages.

Definition. If L_1, L_2 are languages, the quotient L_1/L_2 is defined by

$$L_1/L_2 = \{u \mid \text{there exists } v \in L_2 \text{ such that } uv \in L_1\}.$$

It is true that, if L is deterministic and R is regular, then L/R is deterministic. This is not easy and involves the construction of a "predicting machine". See [20, Theorem 12.4] or [21, Theorem 10.2]. However, we shall only need a special case, which is much easier.

In the special case that $L_2 = \{a\}$, where a is a letter, we write L_1/a, that is, $L_1/a = \{u \mid ua \in L_1\}$. Note that, if L_1 has alphabet A and $a \notin A$, then L_1/a is empty.

Lemma 4.22. *If* L *is deterministic and* a_0 *is any letter, then* L/a_0 *is deterministic.*

Proof. Let $M = (Q, F, A, \Gamma, \tau, q_0, z_0)$ be a deterministic PDA recognising L by final state. Let B be the set of all pairs $(q, z) \in Q \times \Gamma$ such that there is a transition in τ of the form (q, a_0, z, p, α), where $p \in F$. Then for $w \in A^*$, $w \in L/a_0$ if and only if

$$(q_0, w, z_0) \xrightarrow[M]{} (q, a_0, z\gamma)$$

for some $(q, z) \in B$ and $\gamma \in \Gamma^*$. Define a new PDA $M' = (Q', F', A, \Gamma, \tau', q_0, z_0)$ as follows. For every $q \in Q$, take two new states q', q'' and let $Q' = \{q, q', q'' \mid q \in Q\}$. Then put $F' = \{q'' \mid q \in Q\}$. The set τ' is obtained from τ as follows. First, replace every $(q, a, z, p, \alpha) \in \tau$ by (q, a, z, p', α). Then add new transitions as follows:

$$(q', \varepsilon, z, q, z) \qquad \text{for } (q, z) \in Q \times Z, \ (q, z) \notin B$$

$$\left. \begin{array}{l} (q', \varepsilon, z, q'', z) \\ (q'', \varepsilon, z, q, z) \end{array} \right\} \quad \text{for } (q, z) \in B.$$

Clearly $w \in L(M')$ if and only if $(q_0, w, z_0) \xrightarrow[M']{} (q'', \varepsilon, \alpha)$ for some $q \in Q$ and $\alpha \in \Gamma^*$, if and only if $(q_0, w, z_0) \xrightarrow[M']{} (q', \varepsilon, z\gamma)$ for some $(q, z) \in B$ and $\gamma \in \Gamma^*$. Finally, it is easily seen that this is true if and only if $(q_0, w, z_0) \xrightarrow[M]{} (q, \varepsilon, z\gamma)$ for some $(q, z) \in B$ and $\gamma \in \Gamma^*$. Thus $L(M') = L/a_0$, and M' is obviously deterministic. □

Note that, in the proof, if a_0 is not in A, B is empty, so M' recognises the empty language, as it never reaches a final state. This accords with the remark above that L/a_0 is empty.

Theorem 4.23. *Let \$ be a letter not in the alphabet of a language L. If $L\$ = L_G$ for some $LR(0)$ grammar G, then $L = L_{G'}$ for some $LR(1)$ grammar G'.*

Proof. Let $G = (V_N, V_T, P, S)$. Using Lemma 4.3, we can assume that all variables of G are generating. (The construction removes some of the variables and productions, which still leaves an $LR(0)$ grammar.) Construct a grammar $G' = (V_N', V_T', P', S)$ by making \$ a variable rather than a terminal, and adding the production $\$ \longrightarrow \varepsilon$ to P. Thus $V_N' = V_N \cup \{\$\}$, $V_T' = V_T \setminus \{\$\}$ and $P' = P \cup \{\$ \longrightarrow \varepsilon\}$. We shall show G' is $LR(1)$ and $L_{G'} = L$. Since S does not occur on the right of any production of G, it does not occur on the right of any production of G'.

Given a G' derivation of α from S, if there are n uses of $\$ \longrightarrow \varepsilon$, then omitting them gives a G-derivation of a word with at least n occurrences of \$. Since every variable of G is generating, the derivation can be continued to obtain a G-derivation of a word w in V_T^*, still with at least n occurrences of \$, since $\$ \in V_T$. But $w \in L_G = L\$$, so $n \leq 1$.

By Lemma 4.2 and its proof, the G' derivation of α from S defines an S-tree, and the tree defines a rightmost derivation of α from S, using the same productions as the original derivation, but in a possibly different order. (Further, every rightmost derivation is obtained in this way.) This rightmost derivation therefore uses $\$ \longrightarrow \varepsilon$ at most once. Since the derivation is rightmost, if $\$ \longrightarrow \varepsilon$ is used it must remove an occurrence of \$ at the right-hand end of the word. Otherwise, the procedure of the preceding paragraph would give a G-derivation of a word $u\$v \in V_T^*$ with $v \neq \varepsilon$, which is impossible as $L_G = L\$$.

If $\alpha \in (V_T')^*$, then there must be a use of $\$ \longrightarrow \varepsilon$ in the derivation. Otherwise, the derivation is a G-derivation with $\alpha \in (V_T)^*$, but $\alpha \notin L\$$ since $\$ \notin V_T'$, a contradiction. Thus if the G'-derivation is $S = \alpha_0, \alpha_1, \ldots, \alpha_n = \alpha$, then for some i, $\alpha_i = \gamma\$$ and $\alpha_{i+1} = \gamma$. Further, none of $\alpha_0, \ldots, \alpha_{i-1}$ end with $\$$, since the derivation is rightmost. Omitting the use of $\$ \longrightarrow \varepsilon$ gives a G-derivation

$$S, \alpha_1, \ldots, \alpha_i = \alpha_{i+1}\$, \alpha_{i+2}\$, \ldots, \alpha_n\$ = \alpha\$.$$

Consequently, $\alpha\$ \in L\$$ (as $V_T' \subseteq V_T$), so $\alpha \in L$, hence $L_{G'} \subseteq L$. Also, this G'-derivation is rightmost. The only place at which a production $\$ \longrightarrow \varepsilon$ can be inserted into this derivation to get a rightmost G'-derivation is after α_i, giving the original G'-derivation (α_i is the first word in the derivation to end with $\$$). Thus different rightmost G'-derivations of a word $\alpha \in (V_T')^*$ give different rightmost G-derivations of $\alpha\$$. Since G is unambiguous (Lemma 4.20), so is G'.

If $w \in L$, then $S \xrightarrow[G]{\bullet} w\$$ by some G-derivation. Using the G'-production $\$ \longrightarrow \varepsilon$ then gives a G'-derivation of w, so $S \xrightarrow[G']{\bullet} w$, hence $w \in L_{G'}$. Thus $L = L_{G'}$.

If $A \in V_N$, A is generating in G, and by use of $\$ \longrightarrow \varepsilon$, we see that $A \xrightarrow[G']{\bullet} w$ for some $w \in (V_T')^*$, so A is generating in G'. Clearly $\$$ is generating in G' ($\varepsilon \in (V_T')^*$), so all variables of G' are generating.

Since $\$$ has become a variable, to show G' is $LR(1)$, we need to choose a new letter not in $V_N' \cup V_T' = V_N \cup V_T$, which we denote by €. Thus, we have to show that, given rightmost G'-derivations

$$S€, \ldots, \alpha A w_1 w_2, \alpha \beta w_1 w_2$$
$$S€, \ldots, \gamma B w, \alpha \beta w_1 w_3$$

where $|w_1| = 1$, then $\gamma = \alpha$, $A = B$, and $w = w_1 w_3$. There are two possible cases.

(a) $w_1 = €$, in which case $w_2 = w_3 = \varepsilon$.
(b) w_2, w_3 both end in €, say $w_2 = w_2'€$, $w_3 = w_3'€$.

In both cases, w ends in €, say $w = w'€$.
Case (a). Omitting the occurrences of € gives G'-derivations

$$S, \ldots, \alpha A, \alpha \beta$$
$$S, \ldots, \gamma B w', \alpha \beta$$

Since every variable of G' is generating, there is a G'-derivation from $\alpha\beta$ of some $u \in (V_T')^*$, and we can take the derivation to be rightmost (by Lemma 4.2). Adding this derivation to the right of the two derivations of $\alpha\beta$ gives two rightmost G'-derivations of u from S. Since G' is unambiguous, these two derivations are the same, hence the two derivations of $\alpha\beta$ are the same. Consequently, $\alpha A = \gamma B w'$. Since B is the rightmost variable in $\gamma Bw'$, $w' = \varepsilon$, $B = A$ and $\gamma = \alpha$. Hence $w = w'\text{\large€} = w_1 w_3$, as required.

Case (b). In this case, omitting the occurrences of $\text{\large€}$ gives derivations

$$S, \dots, \alpha A w_1 w_2', \alpha \beta w_1 w_2'$$
$$S, \dots, \gamma B w', \alpha \beta w_1 w_3'.$$

Since $w_1 \neq \varepsilon$, the final productions used in these derivations are not $\$ \longrightarrow \varepsilon$. Therefore omitting the single use of this production, if it occurs, from the derivations gives rightmost G-derivations

$$S, \dots, \alpha A w_2'', \alpha \beta w_2''$$
$$S, \dots, \gamma B w'', \alpha \beta w_3''.$$

where w_2'' is either $w_1 w_2'$ or $w_1 w_2' \$$, and either $w'' = w'$, $w_3'' = w_1 w_3'$, or $w'' = w' \$$, $w_3'' = w_1 w_3' \$$. Since G is $LR(0)$, $A = B$, $\gamma = \alpha$ and $w_3'' = w''$. It follows that $w_1 w_3' = w'$, hence $w_1 w_3 = w$, as claimed. □

The next theorem needs a lemma whose proof is quite subtle, and which depends on another non-trivial lemma. The proofs of these two lemmas (A.3 and A.4) have been placed in Appendix A.

Theorem 4.24. *For a language L, the following are equivalent.*

(1) $L = L_G$ *for some k and $LR(k)$ grammar G.*
(2) L *is deterministic.*
(3) $L = L_G$ *for some $LR(1)$ grammar G.*

Proof. Again let $\$$ be a letter not in the alphabet of L.

Assume (1). By Theorem 4.18, $L\$^k$ is deterministic. For $k > 0$, $L\$^{k-1} = L\$^k / \$$, and by an easy induction on k, using Lemma 4.22, L is deterministic, so (2) holds.

Assume (2). Then $L\$$ is deterministic. For $L = L(M)$ for some deterministic PDA $M = (Q, F, A, \Gamma, \tau, q_0, z_0)$ and by Lemma A.4, we can assume M has no transitions starting with $(q, \varepsilon, \dots)$, where $q \in F$. Let $M' = (Q \cup \{f\}, \{f\}, A \cup \{\$\}, \Gamma, \tau', q_0, z_0)$ where the transitions in τ' are those in τ, together with $(q, \$, z, f, z)$ for all $q \in F$ and $z \in \Gamma$. Then M' is deterministic and it is easy to see that $L\$ = L(M')$.

Clearly $L\$$ is prefix-free, hence $L\$$ is L_G for some $LR(0)$ grammar G by Theorem 4.21. Now (3) follows by Theorem 4.23. Obviously (3) implies (1). □

The deterministic and strict deterministic languages can also be characterised by what are called deterministic and strict deterministic grammars. See [12], §11.4 and §11.8, Problem 4. Note however, that a different definition of $LR(k)$ is used in [12]. This gives a different class of languages generated by $LR(0)$ grammars (see [12, Theorem 13.3.1]). For further discussion, see the problems at the end of [12, §13.2].

We now give some examples to clarify the inclusion relations between the classes of languages we have studied.

Examples.

(1) (Example 10.1 in [21].) $L = \{0^i 1^j 2^k \mid i = j \text{ or } j = k\}$ is context-free, being generated by the grammar with $V_N = \{A, B, C, D, S\}$, $V_T = \{0, 1, 2\}$ and productions

$$S \longrightarrow AB \mid CD, \; A \longrightarrow 0A1 \mid \varepsilon, \; B \longrightarrow 2B \mid \varepsilon, \; C \longrightarrow 0C \mid \varepsilon, \; D \longrightarrow 1D2 \mid \varepsilon.$$

But L is not deterministic. Otherwise L^c would be deterministic (see the note after Lemma A.4 in Appendix A), so context-free. The language $0^* 1^* 2^*$ (meaning $\{0\}^* \{1\}^* \{2\}^*$) is regular by Lemma 1.5, so $L_1 := L^c \cap 0^* 1^* 2^*$ is context-free by Lemma 4.16. But $L_1 = \{0^i 1^j 2^k \mid i \neq j \text{ and } j \neq k\}$, which is not context-free, by a generalisation of the Pumping Lemma due to Ogden ([21, Lemma 6.2]). Other examples of context-free, non-deterministic languages, given in §6.4 of [22], are

$$\{0^n 1^n \mid n \geq 1\} \cup \{0^n 1^{2n} \mid n \geq 1\}$$

and the set of even-length palindromes on the alphabet $\{0, 1\}$.

(2) The language $L = \{w \in \{a, b\}^* \mid w$ has an equal number of a's and b's$\}$ is deterministic. It is left as an exercise to construct a deterministic PDA recognising L by final state. However, it is not prefix-free, so is not strict deterministic. Also, L is not regular. For suppose it is. Choose p as in the Pumping Lemma (Lemma 1.8). Let $x = a^p b^p$, and decompose x as uvw as in this lemma. Since $|uv| \leq p$, uv consists entirely of a's. Taking $i = 0$ in the Pumping Lemma, $uw \in L$. But all b's in x occur in w, and the number of a's in uw is less than p, since $v \neq \varepsilon$. Hence uw has more b's than a's, so $uw \notin L$, a contradiction.

(3) The language $\{0^n 1^n \mid n \geq 1\}$ is strict deterministic (this is left as an exercise), but is not regular (see the example after Theorem 1.7, or use the Pumping Lemma as in Example 2).

(4) The language $\{0^n \mid n \geq 1\}$ is regular (see Example 1 near the beginning of Chapter 1) but is not prefix-free, so is not strict deterministic.

Using these and examples from previous chapters, together with some of the results which have been proved, there is thus a hierarchy of language classes as illustrated below, where a class is strictly contained in a class above joined to it by a line.

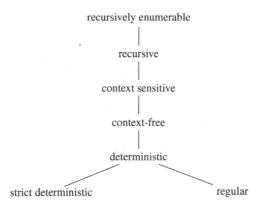

Figure 4.5

We remark that the diagram can be considerably elaborated, in particular by some of the classes mentioned at the end of Chapter 3, for suitable choices of $f(n)$. (These classes are known as *complexity classes*). Although there is no inclusion relation between the bottom two classes, they do intersect. A regular, prefix-free language L is deterministic, so L is strict deterministic, by Theorem 4.21. A simple example is $L = \{w\}$, where w is a non-empty word. More elaborate examples can be found in the exercises for §2.2 in [21].

Exercises on Chapter 4

1. Find a grammar in Chomsky Normal Form generating the same language as the grammar

$$G = (\{A, B, S\}, \{a, b\}, P, S)$$

where P consists of the productions

$$S \longrightarrow AA \mid B$$
$$A \longrightarrow aA \mid B \mid BBB$$
$$B \longrightarrow b$$

2. Find a grammar in Greibach Normal Form generating the same language as the grammar

$$G = (\{A, B, S\}, \{a, b\}, P, S)$$

where P consists of the productions

$$S \longrightarrow SA \mid a$$
$$A \longrightarrow B \mid a$$
$$B \longrightarrow Ab$$

3. Let $G = (V_N, V_T, P, S)$ be a context-free grammar, and suppose P contains a production $A \longrightarrow uv$, where $u, v \in (V_N \cup V_T)^*$. Let G' be obtained by adding a new variable C and replacing $A \longrightarrow uv$ by the two productions $A \longrightarrow uC$ and $C \longrightarrow v$. Show that $L_G = L_{G'}$. If, instead, we replace $A \longrightarrow uv$ by $A \rightarrow Cv$ and $C \longrightarrow u$, show that the language generated is not changed.

4. A grammar is said to be *linear* if all productions are of the form $A \rightarrow uBv$ or $A \rightarrow u$, where A, B are variables and u, v are strings of terminals (possibly empty). (Thus a linear grammar is context-free, and regular grammars are linear.) A language is *linear* if it is generated by a linear grammar.

(a) If L is a linear language, show that L is generated by a grammar with all productions of the form $A \longrightarrow uB$, $A \longrightarrow Bu$ or $A \longrightarrow u$, where A, B are variables and u is a string of terminals.

(b) If L is linear, show that $L \setminus \{\varepsilon\}$ is generated by a grammar with all productions of the form $A \longrightarrow aB$, $A \longrightarrow Ba$ or $A \longrightarrow a$, where A, B are variables and a is a terminal. (Hint: see Remark 4.4.)

(c) Give a grammar in the form of part (b) generating $\{0^n 1^n \mid n > 0\}$. (It is linear but not regular-see the example after Theorem 1.7. It may help to start with the grammar in Example (3), p.3.)

5. Prove the Pumping Lemma for linear languages. *Let L be a linear language. Then there is an integer $p > 0$, depending only on L, such that, if $z \in L$ and $|z| \geq p$, then z can be written as $z = uvwxy$, where $|uvxy| \leq p$, v and x are not both ε and for every $i \geq 0$, $uv^i wx^i y \in L$.* [Hint: consider parsing trees for a grammar generating $L \setminus \{\varepsilon\}$ in the form of Exercise 4(b). Argue as in the proof of the Pumping Lemma for context-free languages; p, and the vertices v_r, v_s need to be chosen differently.]

6. Show that $\{0^m 1^m 0^n 1^n \mid m, n > 0\}$ is context-free but not linear.

Chapter 5
Connections with Group Theory

There have been connections between formal language theory and group theory for a long time. The original connections involved certain decision problems, and we shall study one of these, the word problem. Given a group G and a finite set of generators, this asks if there is a procedure with a finite set of instructions to determine whether or not a word in the generators and their inverses represents 1 in G. The set W of words representing 1 in G is a language, so the question is whether or not W is decidable. The formal version of the word problem therefore asks whether or not W is recursive. (The answer is no, in general.) We prove the result of Anisimov, that W is regular if and only if G is finite. We also prove that W is context-free if and only if G has a free subgroup of finite index. The proof is not self-contained as it uses results of Dunwoody on accessible groups, and results of Gregorac and of Karrass, Pietrowski and Solitar are quoted. The part of the proof we give is due to Muller and Schupp and is the heart of the proof. We finish with a brief look at automatic groups; these form an interesting class of groups which has been well studied recently. We prove the characterisation of automatic groups by means of the "fellow traveller" property in the *Cayley graph*, a graph associated with a set of generators of the group. We begin with some discussion of group presentations and free groups.

Presentations of Groups

Let X be a set of generators of a group G. If $f, g : G \longrightarrow H$ are two homomorphisms which agree on X, then $f = g$. (The *equaliser* $\{a \in G \mid f(a) = h(a)\}$ is a subgroup of G, and contains X, so equals G). However, given a mapping $f : X \longrightarrow H$, there is no guarantee that f extends to a homomorphism from G to H. As a simple example, let G be cyclic of order 2 generated by x and let H be cyclic of order 3, generated by y. Then there is no homomorphism $f : G \longrightarrow H$ such that $f(x) = y$, because $x^2 = 1$, but $y^2 \neq 1$. We begin by investigating when an extension to a homomorphism exists.

   Since X generates G, every element of G can be expressed as $x_1^{e_1} \ldots x_n^{e_n}$, where $x_i \in X$, $e_i = \pm 1$, $n \geq 0$. There are many different ways of expressing a given element of G in this form. When we say "different ways", we are viewing $x_1^{e_1} \ldots x_n^{e_n}$ as a string, rather than a product of elements of G. Also, it may happen that $x = x^{-1}$ for

I. Chiswell, *A Course in Formal Languages, Automata and Groups*,
DOI 10.1007/978-1-84800-940-0_5,
© Springer-Verlag London Limited 2009

some $x \in X$, and to express that the strings x, x^{-1} represent the same element of G, we cannot view x^{-1} as the inverse of x in G.

In view of this, we proceed as follows. Let X be a set, and let X^{-1} be a set, in one-to-one correspondence with X via a mapping $x \mapsto x^{-1}$, and with $X \cap X^{-1} = \emptyset$. Let $X^{\pm 1} = X \cup X^{-1}$. We can extend the mapping to an involution $X^{\pm 1} \to X^{\pm 1}$ without fixed points, by defining $(x^{-1})^{-1} = x$, for $x \in X$. This involution can then be extended to $(X^{\pm 1})^*$ by defining $(y_1 \ldots y_n)^{-1} = (y_n^{-1} \ldots y_1^{-1})$ for $y_i \in X^{\pm 1}$, and $\varepsilon^{-1} = \varepsilon$. Now every element of $(X^{\pm 1})^*$ represents an element of G in an obvious way, and different words in $(X^{\pm 1})^*$ may represent the same element of G. If u, v represent the same element of G, then we say the relation $u = v$ holds in G.

We can start with a set X and write down certain relations, then consider a group G generated by X in which these relations hold. However, such a group G might not exist, because the relations may imply the relation $x = y$ holds in G, where $x, y \in X$ and $x \neq y$. For example, Let $X = \{x, y\}$ and let the relations be $xyx^{-1} = yxy^{-1}$ and $xy = yx$. (There are less obvious examples.)

To cater for this, we instead consider a set X and a mapping (of sets) $\varphi : X \longrightarrow G$, where G is a group. Then $(X^{\pm 1})^*$ is a monoid under concatenation, and φ extends to a monoid homomorphism $\overline{\varphi} : (X^{\pm 1})^* \longrightarrow G$ by

$$\overline{\varphi}(x_1^{e_1} \ldots x_n^{e_n}) = \varphi(x_1)^{e_1} \ldots \varphi(x_n)^{e_n}$$

$(x_i \in X, e_i = \pm 1)$, and $\overline{\varphi}(\varepsilon) = 1_G$ (the identity element of G). Also, $\overline{\varphi}(u^{-1}) = \overline{\varphi}(u)^{-1}$ for $u \in (X^{\pm 1})^*$. Note that $\overline{\varphi}$ is surjective if and only if $\varphi(X)$ generates G.

Let $u, v \in (X^{\pm 1})^*$. We say that the relation $u = v$ holds in G (via φ) if u, v represent the same element of G, that is, $\overline{\varphi}(u) = \overline{\varphi}(v)$. Formally, a relation on X is an ordered pair of words over the alphabet $(X^{\pm 1})^*$, but we always speak of the relation $u = v$ rather than (u, v). Note that, if $v = \varepsilon$, the relation is written $u = 1$, and similarly if $u = \varepsilon$. (We shall show in Lemma 5.3 below that this works; given a set X and certain relations, there is a group G and mapping $\varphi : X \longrightarrow G$ such that $\varphi(X)$ generates G and these relations hold in G via φ.) We can give a criterion for extension of homomorphisms, now complicated by the presence of the mapping φ.

Lemma 5.1. *Let $\varphi : X \longrightarrow G$, $\alpha : X \longrightarrow H$ be maps of sets, where G, H are groups. Suppose $\varphi(X)$ generates G. Then there is a homomorphism $\widetilde{\alpha} : G \longrightarrow H$ such that $\widetilde{\alpha}\varphi = \alpha$ if and only if, for all relations $u = v$,*

($*$) $u=v$ *holds in G (via φ) implies $u=v$ holds in H (via α)*

Proof. Statement ($*$) is equivalent to: $\overline{\varphi}(u) = \overline{\varphi}(v)$ implies $\overline{\alpha}(u) = \overline{\alpha}(v)$, for all u, $v \in (X^{\pm 1})^*$. Also, $\varphi(X)$ generates G if and only if $G = \overline{\varphi}((X^{\pm 1})^*)$.

Thus, if ($*$) is satisfied, we can define $\widetilde{\alpha}$ by $\widetilde{\alpha}(\overline{\varphi}(u)) = \overline{\alpha}(u)$. It is easily checked that $\widetilde{\alpha}$ is a homomorphism, and clearly $\widetilde{\alpha}\varphi = \alpha$.

Conversely, if $\widetilde{\alpha}$ exists, and $\overline{\varphi}(u) = \overline{\varphi}(v)$, then $\widetilde{\alpha}(\overline{\varphi}(u)) = \widetilde{\alpha}(\overline{\varphi}(v))$. But since $\widetilde{\alpha}$ is a homomorphism and $\widetilde{\alpha}\varphi = \alpha$, $\widetilde{\alpha}\overline{\varphi} = \overline{\alpha}$, hence $\overline{\alpha}(u) = \overline{\alpha}(v)$. $\square$

If R is a set of relations, we say that R *holds in* G (*via* φ) if every element of R holds in G via φ. We have already used the idea that certain relations can imply others, and this can be formalised as follows.

Definition. A relation $u = v$ is a *consequence* of R if, for all groups H and maps $\alpha : X \longrightarrow H$,

$$\text{if } R \text{ holds in } H \text{ via } \alpha \text{ then } u = v \text{ holds in } H \text{ via } \alpha.$$

To see that the formal definition of "consequence" captures the idea of relations in a group implying another relation, consider an example: $xy = yx$ is a consequence of $\{x^2 = 1, y^2 = 1, (xy)^2 = 1\}$. For a mapping $\alpha : \{x, y\} \to H$ corresponds to a choice of two elements $a = \alpha(x)$ and $b = \alpha(y)$, so the assertion is that, if a and b are any elements of a group H satisfying $a^2 = b^2 = (ab)^2 = 1$, then $ab = ba$, which is easy to see. (In practice, one usually suppresses the mapping α and just observes that if x, y are group elements satisfying $x^2 = 1$, $y^2 = 1$, $(xy)^2 = 1$ then $xy = yx$.)

Definition. A group presentation consists of a set X and a set R of relations on X, denoted by $\langle X \mid R \rangle$.

Let $\varphi : X \longrightarrow G$ be a mapping, where G is a group. The presentation $\langle X \mid R \rangle$ is called a *presentation of* G (*via* φ) if $\varphi(X)$ generates G, and a relation holds in G via φ if and only if it is a consequence of R. In these circumstances, R is called a set of *defining relations* for G.

Concerning notation, if $R = \{u_1 = v_1, \ldots, u_n = v_n\}$, the presentation is written

$$\langle X \mid u_1 = v_1, \ldots, u_n = v_n \rangle.$$

If $R = \{u_i = v_i \mid i \in I\}$ is an indexed set, we write $\langle X \mid u_i = v_i \ (i \in I) \rangle$. Similar conventions apply to X.

We write $G = \langle X \mid R \rangle^{\varphi}$ to mean G has presentation $\langle X \mid R \rangle$ via φ. Clearly, $u = v$ holds in G if and only if $uv^{-1} = 1$ holds in G. It follows that if R' is obtained by replacing some or all of the relations $u = v$ in R by $uv^{-1} = 1$, then $G = \langle X \mid R' \rangle^{\varphi}$. Thus we can assume, if necessary, that all elements of R have the form $u = 1$.

As a simple example, let $X = \{x\}$ and let φ map x to a generator of a cyclic group of order n, where n is a positive integer. If $u \in (X^{\pm 1})^*$, then by deleting pairs xx^{-1} or $x^{-1}x$, we obtain a word $v = x^k$, where $k \in \mathbb{Z}$, such that $u = 1$ is a consequence of $v = 1$ and vice-versa. Then $v = 1$ holds via φ if and only if n divides k, in which case $v = 1$ is a consequence of $x^n = 1$. Hence the relations which hold via φ are precisely the consequences of $x^n = 1$. Therefore, $\langle x \mid x^n = 1 \rangle$ is a presentation of the cyclic group of order n.

Lemma 5.2. *Let $G = \langle X \mid R \rangle^{\varphi}$ and let $\alpha : X \longrightarrow H$ be a mapping, where H is a group. Then the following are equivalent:*

(1) *R holds in H via α.*
(2) *there is a unique homomorphism $\widetilde{\alpha} : G \longrightarrow H$ such that $\widetilde{\alpha}\varphi = \alpha$;*

Proof. Assume (1). If $u = v$ holds in G, it is a consequence of R, so holds in H, and $\tilde{\alpha}$ as in (2) exists by Lemma 5.1. Uniqueness follows because $\varphi(X)$ generates G.

Assume (2). Then R holds in G via φ and $\tilde{\alpha}\overline{\varphi} = \overline{\alpha}$ (all the maps are monoid homomorphisms preserving inverses). Hence R holds in H (via α). $\square$

The observation in the proof, that if (2) holds then $\tilde{\alpha}\overline{\varphi} = \overline{\alpha}$, should be borne in mind. If, in Lemma 5.2, $H = \langle X \mid R \rangle^{\alpha}$, then (1) is satisfied, and the mapping $\tilde{\alpha}$ given by (2) is an isomorphism. For if $\overline{\varphi}(u) \in \mathrm{Ker}(\alpha)$, $\tilde{\alpha}(\overline{\varphi}(u)) = 1 = \overline{\alpha}(u)$. Hence the relation $u = 1$ holds in H, so is a consequence of R, so holds in G, that is, $\overline{\varphi}(u) = 1$. Thus $\tilde{\alpha}$ is injective. It is surjective as $\overline{\alpha}(X)$ generates H. Thus if two groups have the same presentation, via possibly different maps, they are isomorphic.

On the other hand, if $G = \langle X \mid R \rangle^{\varphi}$ and $f : G \longrightarrow H$ is an isomorphism, then $H = \langle X \mid R \rangle^{f\varphi}$. (It is easy to see that a relation holds in H via $f\varphi$ if and only if it holds in G via φ.)

Lemma 5.3. *If $\langle X \mid R \rangle$ is a group presentation, then there exist a group G and a mapping $\varphi : X \longrightarrow G$ such that $G = \langle X \mid R \rangle^{\varphi}$.*

Proof. For $u, v \in (X^{\pm 1})^*$, define $u \equiv_R v$ to mean $u \xrightarrow{\;*\;}_P v$, where P is the set containing the following productions.

(1) $r \longrightarrow s, r^{-1} \longrightarrow s^{-1}, s \longrightarrow r$ and $s^{-1} \longrightarrow r^{-1}$, for all relations $r = s$ in R.
(2) $yy^{-1} \longrightarrow \varepsilon$ and $\varepsilon \longrightarrow yy^{-1}$, for all $y \in X^{\pm 1}$.

It is easily checked that $\equiv_R$ is an equivalence relation on $(X^{\pm 1})^*$. Let $[u]$ (or $[u]_R$ if necessary) denote the equivalence class of $u \in (X^{\pm 1})^*$. If $u_1, \ldots, u_k$ is a P-derivation, so is $u_1 w, \ldots, u_k w$, for any $w \in (X^{\pm 1})^*$, so $u \equiv_R v$ implies $uw \equiv vw$, and similarly $u \equiv_R v$ implies $wu \equiv wv$. Hence, if $u \equiv_R u'$ and $v \equiv_R v'$, then $uv \equiv_R uv' \equiv_R u'v'$, so $uv \equiv_R u'v'$. We can therefore define a binary operation on $(X^{\pm 1})^*/ \equiv_R$ by $[u][v] = [uv]$. This makes $(X^{\pm 1})^*/ \equiv_R$ into a group, which we denote by G. The identity element is $[\varepsilon]$, and $[u]^{-1}$ is $[u^{-1}]$.

Define $\varphi : X \longrightarrow G$ by $\varphi(x) = [x]$. To show $G = \langle X \mid R \rangle^{\varphi}$, we have to show that, for any words u, v, $\overline{\varphi}(u) = \overline{\varphi}(v)$ if and only if $u = v$ is a consequence of R.

Suppose $\overline{\varphi}(u) = \overline{\varphi}(v)$ and $\alpha : X \longrightarrow H$ is a map, where H is a group and R holds in H via α. It is easily seen that $\overline{\varphi}(u) = [u]$ for all $u \in (X^{\pm 1})^*$, so $u \equiv_R v$. Therefore there is a P-derivation $u = u_1, \ldots, u_k = v$, and it follows by induction on k that $\overline{\alpha}(u) = \overline{\alpha}(v) = \overline{\alpha}(u_i)$ for $1 \leq i \leq k$, hence $u = v$ is a consequence of R. One has to check several (easy) cases, for example $u_{k-1} = w_1 r w_2$, $u_k = w_1 s w_2$, where one of $r = s, r^{-1} = s^{-1}, s = r, s^{-1} = r^{-1}$ is in R. Then $\overline{\alpha}(r) = \overline{\alpha}(s)$, hence $\overline{\alpha}(u_{k-1}) = \overline{\alpha}(u_k)$. The remaining cases are left to the reader.

Conversely, suppose $u = v$ is a consequence of R. If $r = s$ is a relation in R, then $r \equiv_R s$ via a derivation with one step, using the production $r \longrightarrow s$. Hence $\overline{\varphi}(r) = \overline{\varphi}(s)$, so R holds in G via φ. By assumption, $\overline{\varphi}(u) = \overline{\varphi}(v)$. $\square$

Example. $\langle x, y \mid x^2 = 1, y^3 = 1, xyx^{-1} = y^{-1} \rangle$ is a presentation of S_3, the symmetric group of degree 3, via α, where $\alpha(x) = (1, 2)$, $\alpha(y) = (1, 2, 3)$. For if G has this presentation, via φ, say, then by Lemma 5.2, there is a homomorphism $\tilde{\alpha} : G \longrightarrow S_3$ with $\tilde{\alpha}(\varphi(x)) = (1, 2)$, $\tilde{\alpha}(\varphi(y)) = (1, 2, 3)$, and it suffices to show that $\tilde{\alpha}$ is an isomorphism. It is onto, as $(1, 2)$ and $(1, 2, 3)$ generate S_3, so it suffices to show $|G| \leq 6$.

Let H be the subgroup of G generated by y (suppressing φ), so $|H| \leq 3$. The set $\{H, Hx\}$ is invariant under right translation by x and y ($Hxy = Hy^{-1}x = Hx$), so by all elements of G. Since the action of G on the right cosets of H by right translation is transitive, this set is the set of all right cosets of H in G, so $(G : H) \leq 2$. Hence $|G| \leq 6$, as required.

The example illustrates the point that the mapping φ, although strictly necessary for the theory, is very often omitted in practice, to keep the notation simple.

Given any group G and mapping $\varphi : X \longrightarrow G$ such that $\varphi(X)$ generates G, let R be the set of all relations holding in G via φ. Then $G = \langle X \mid R \rangle^{\varphi}$, so any group has a presentation. One possibility is to take $X = G$ and φ to be the inclusion map. In this case we can find a smaller set of relations, as follows. Let G be a group, and take a set X in 1-1 correspondence with G, via a mapping $g \mapsto x_g$, for $g \in G$. (This is to avoid confusion between concatenation of words and product in G.) Let $\varphi : X \longrightarrow G$ be the inverse mapping $x_g \mapsto g$, and let R be the set of relations $\{x_g x_h = x_{gh} \mid g, h \in G\}$. We claim that $G = \langle X \mid R \rangle^{\varphi}$. Clearly $\varphi(X) = G$ and R holds in G via φ. Suppose $u = v$ holds in G, and $\alpha : X \longrightarrow H$ is a mapping such that R holds in H via α. To finish the proof, we have to show that $\overline{\alpha}(u) = \overline{\alpha}(v)$. Now, since R holds in H, $\alpha(x_1)\alpha(x_1) = \overline{\alpha}(x_1 x_1)) = \alpha(x_1)$ (as $x_1 x_1 = x_1$ is in R). Hence $\alpha(x_1) = 1_H$. Similarly, as $x_g x_{g^{-1}} = x_1$ is in R, $\overline{\alpha}(x_g^{-1}) = \alpha(x_{g^{-1}})$. For similar reasons, $\alpha(x_g)\alpha(x_h) = \alpha(x_{gh})$. We can also replace α by φ in these formulas.

Write $u = x_{g_1}^{\pm 1} \ldots x_{g_n}^{\pm 1}$; by induction on n, we obtain $\overline{\alpha}(u) = \overline{\alpha}(x_{g_1 \pm 1 \ldots g_n \pm 1})$. Similarly,

$$\overline{\varphi}(u) = \overline{\varphi}(x_{g_1 \pm 1 \ldots g_n \pm 1}) = g_1^{\pm 1} \ldots g_n^{\pm 1}.$$

Thus $\overline{\alpha}(u) = \alpha(x_{\overline{\varphi}(u)})$, and similarly $\overline{\alpha}(v) = \alpha(x_{\overline{\varphi}(v)})$. Since $\overline{\varphi}(u) = \overline{\varphi}(v)$ by assumption, $\overline{\alpha}(u) = \overline{\alpha}(v)$. This presentation of G is called the *standard presentation* of G (or *multiplication table presentation* of G) and is denoted by $\langle G \mid \text{rel } G \rangle$.

An important special case in Lemma 5.3 is when R is empty. The corresponding group $(X^{\pm 1})^* / \equiv_{\emptyset}$ in the proof is called the *free group on X*, denoted by $F(X)$.

Definition. An element of $(X^{\pm 1})^*$ is *reduced* if it has no subword yy^{-1}, where $y \in X^{\pm 1}$.

Lemma 5.4. (Normal Form Theorem) *Every element of $F(X)$ is $[u]_{\emptyset}$ for a unique reduced word u. In particular, X embeds in $F(X)$ via $x \mapsto [x]_{\emptyset}$.*

Proof. In this case, P in the proof of Lemma 5.3 only contains the productions (2). Using the productions $yy^{-1} \longrightarrow \varepsilon$, it is easy to see that every element of $F(X)$ is $[u]$ for some reduced word u.

Suppose $[u] = [v]$, where u, v are reduced, so there is a P-derivation $u = u_1, \ldots, u_k = v$. To prove $u = v$, it suffices to show that, if $k \geq 2$, this G-derivation can be shortened. For then by repeated use of this fact, we can obtain a derivation with $k = 1$, so $u = v$. Note that $k \neq 2$ as u, v are reduced.

Suppose $k > 2$, and let u_i be a word of maximal length in the derivation. Then $1 < i < k$ since u, v are reduced. Further, u_i is obtained from u_{i-1} by inserting yy^{-1} for some $y \in X^{\pm 1}$, and u_{i+1} is obtained from u_i by deleting zz^{-1} for some $z \in X^{\pm 1}$.

If the subwords yy^{-1} and zz^{-1} of u_i coincide or overlap by a single letter, then $u_{i-1} = u_{i+1}$, and u_i, u_{i+1} can be omitted from the derivation.

Otherwise, we can replace u_i by u'_i, where u'_i is obtained from u_i by deleting zz^{-1}, and u_{i+1} is obtained from u'_i by inserting yy^{-1}. This reduces $\sum_{i=1}^{k} |u_i|$, so after finitely many such replacements we shall be able to shorten the derivation. $\square$

In view of this, we identify x with $[x]_\emptyset$, for $x \in X$. The next result is the "universal mapping property" of a free group.

Lemma 5.5. *If $\alpha : X \to H$ is a map, where X is a set and H is any group, there is a unique extension to a homomorphism $\widetilde{\alpha} : F(X) \to H$, given by $[u]_\emptyset \mapsto \overline{\alpha}(u)$.*

Proof. This is immediate from Lemma 5.2 (remember that, in Lemma 5.2(2), $\widetilde{\alpha}\overline{\varphi} = \overline{\alpha}$). $\square$

Suppose R is a set of relations on X which are all of the form $r = 1$. We can just write r instead of $r = 1$ for the elements of R, so R is viewed as a subset of $(X^{\pm 1})^*$, and we say that a relation is a consequence of R, rather than of $\{r = 1 \mid r \in R\}$. The elements of R are then called *relators*. We shall also (inaccurately) not distinguish u and $[u]_\emptyset$, so R is viewed as a subset of $F(X)$. Thus in Lemma 5.5, we now write $\widetilde{\alpha}(u) = \overline{\alpha}(u)$. With this in mind, we can state the next lemma. First, recall that if S is a subset of a group G, the normal subgroup of G generated by S (or normal closure of S in G) is the intersection of all normal subgroups of G containing S, so the smallest normal subgroup containing S. It is the subgroup $\langle S^G \rangle$ generated by S^G, the set of all conjugates of elements of S in G.

Lemma 5.6. *In the previous lemma, let R be a subset of $(X^{\pm 1})^*$. Then*

$$u = v \text{ is a consequence of } R \text{ if and only if } uv^{-1} \in \langle R^{F(X)} \rangle.$$

Proof. Let $N = \langle R^{F(X)} \rangle$. Assume $u = v$ is a consequence of R. Let $\alpha : X \longrightarrow F(X)/N$ be the mapping $x \mapsto xN$, and $\widetilde{\alpha}$ the homomorphism in Lemma 5.5. Then $\overline{\alpha}(r) = \widetilde{\alpha}(r) = rN = 1$ for all $r \in R$, as $R \subseteq N$. Thus the relations $r = 1$ hold in $F(X)/N$ via α, for $r \in R$. Hence $u = v$ holds in $F(X)/N$, so $uv^{-1} = 1$ does, that is, $1 = \overline{\alpha}(uv^{-1}) = uv^{-1}N$, so $uv^{-1} \in N$.

Conversely, assume $uv^{-1} \in N$. Let $\alpha : X \longrightarrow G$ be a mapping such that R holds in G via α. Let $\widetilde{\alpha} : F(X) \longrightarrow G$ be the homomorphism given by Lemma 5.5. Then $\widetilde{\alpha}(r) = \overline{\alpha}(r) = 1$ for $r \in R$, that is, $R \subseteq \text{Ker}(\widetilde{\alpha})$, so $N \subseteq \text{Ker}(\widetilde{\alpha})$. Hence $\widetilde{\alpha}(uv^{-1}) = 1$, so $\widetilde{\alpha}(u) = \widetilde{\alpha}(v)$, that is, $\overline{\alpha}(u) = \overline{\alpha}(v)$. Hence $u = v$ is a consequence of R. $\square$

Corollary 5.7. *In Lemma 5.5, let R be a subset of $(X^{\pm 1})^*$. The following are equivalent.*

(1) $H = \langle X \mid R \rangle^\alpha$;
(2) *R generates $\text{Ker}(\widetilde{\alpha})$ as a normal subgroup of $F(X)$ and $\alpha(X)$ generates H.*

If (1) and (2) hold, then $u = 1$ holds in H via α if and only if we can write

$$u =_{F(X)} \prod_{i=1}^{k} u_i r_i^{e_i} u_i^{-1} \qquad (**)$$

for some $k \in \mathbb{N}$, $u_i \in F(X)$, $r_i \in R$ and $e_i = \pm 1$, where $=_{F(X)}$ means $\equiv_{\emptyset}$.

Proof. Let $N = \langle R^{F(X)} \rangle$. Assume (1). Clearly $R \subseteq \mathrm{Ker}(\widetilde{\alpha})$, so $N \subseteq \mathrm{Ker}(\widetilde{\alpha})$. For the reverse inclusion, suppose $u \in \mathrm{Ker}(\widetilde{\alpha})$. Then $\overline{\alpha}(u) = \widetilde{\alpha}(u) = 1$, so $u = 1$ is a relation holding in H, hence is a consequence of R. By Lemma 5.6, $u \in N$. Thus $N = \mathrm{Ker}(\widetilde{\alpha})$ and (2) follows.

Assume (2), so $N = \mathrm{Ker}(\widetilde{\alpha})$. Then a relation $u = v$ holds in H via α if and only if $\overline{\alpha}(u) = \overline{\alpha}(v)$, if and only if $\overline{\alpha}(uv^{-1}) = \widetilde{\alpha}(uv^{-1}) = 1$, i.e. $uv^{-1} \in N$. By Lemma 5.6, this happens if and only if $u = v$ is a consequence of R, hence (1) holds. In particular, if $u = 1$ holds in H via α then $u \in N$, and the last part of the lemma follows. $\qquad \square$

Consequently, if $H = \langle X \mid R \rangle^{\alpha}$, H is isomorphic to $F(X)/N$, where $N = \langle R^{F(X)} \rangle$. This is often used as an alternative way to define a group with presentation $\langle X \mid R \rangle$.

Note that, when $X = \emptyset$, $F(X)$ is the trivial group, and when X has one element, $F(X)$ is infinite cyclic, by Lemma 5.4. If X has more than one element, $F(X)$ is non-abelian. Any group isomorphic to $F(X)$ for some X is called a *free group*. See the exercises at the end of the chapter for more information. For further theory of free groups, see [25, Chapter I]. One important fact that we shall not prove is the Nielsen-Schreier Theorem, that a subgroup of a free group is a free group. Proofs can be found in [5] and [25].

Free Products with Amalgamation. Suppose $\{G_i \mid i \in I\}$ is a family of groups with a common subgroup A, such that $G_i \cap G_j = A$ for $i \neq j$. The family is then called an *amalgam* of groups. If G is a group containing $\bigcup_{i \in I} G_i$ and each G_i is a subgroup of G, we say that G *embeds* the amalgam. We shall show that such a group G always exists.

Instead of an amalgam, consider a family $\{G_i \mid i \in I\}$ and a family of monomorphisms $\alpha_i : A \longrightarrow G_i$, for some fixed group A. Does there exist a group G and monomorphisms $f_i : G_i \longrightarrow G$ such that $\{f_i(G_i) \mid i \in I\}$ is an amalgam with $f_i \alpha_i$ independent of i and $f_i(G_i) \cap f_j(G_j) = f_i \alpha_i(A)$ for $i \neq j$? The answer is yes. (This implies the result of the previous paragraph, taking the α_i to be inclusion maps.) In fact we shall show that a suitable group G is the "free product of the G_i with A amalgamated", defined by the following universal mapping property.

Definition. Let $\{G_i \mid i \in I\}$ be a family of groups and $\alpha_i : A \to G_i$ a monomorphism, for all $i \in I$. A group G is the *free product of the G_i with A amalgamated* (via the α_i) if there exist homomorphisms $f_i : G_i \longrightarrow G$ such that $f_i \alpha_i = f_j \alpha_j$ for all $i, j \in I$, and if $h_i : G_i \longrightarrow H$ are homomorphisms with $h_i \alpha_i = h_j \alpha_j$ for all $i, j \in I$, then there is a unique homomorphism $h : G \longrightarrow H$ such that $h f_i = h_i$ for all $i \in I$.

This is illustrated by a commutative diagram:

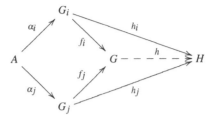

Figure 5.1

We refer to h as an *extension* of the maps h_i.

To establish our claims, we shall show that G exists, that the f_i are monomorphisms and $\{f_i(G_i) \mid i \in I\}$ is an amalgam as above, which G embeds. The uniqueness of G up to isomorphism follows on general category-theoretic grounds (it is a special kind of colimit). Explicitly, let $f_i' : G_i \longrightarrow G'$ be homomorphisms such that $f_i' \alpha_i = f_j' \alpha_j$ for all $i, j \in I$, and if $h_i : G_i \longrightarrow H$ are homomorphisms with $h_i \alpha_i = h_j \alpha_j$ for all $i, j \in I$, then there is a unique homomorphism $h' : G' \longrightarrow H$ such that $h' f_i' = h_i$ for all $i \in I$.

In the definition of G, take $h_i = f_i'$, to obtain a homomorphism $f : G \longrightarrow G'$ such that $f f_i = f_i'$ for all i. Interchanging the roles of G and G', we obtain $f' : G' \longrightarrow G$ such that $f' f_i' = f_i$ for all i. Take $H = G$, $h_i = f_i$ in the definition; both $f' f$ and $\mathrm{id}_G : G \longrightarrow G$ are extensions of the maps $f_i : G_i \longrightarrow G$. Since the extension is unique, $f' f = \mathrm{id}_G$. Interchanging the roles of G and G', $f f' = \mathrm{id}_{G'}$, so f and f' are inverse isomorphisms.

A similar argument shows that G is generated by $\bigcup_{i \in I} f_i(G_i)$. For let G_0 be the subgroup of G generated by this set. Take $h_i = f_i$, viewed as a mapping to G_0, in the definition of G. There is an extension to a homomorphism $f : G \longrightarrow G_0$. Let $\iota : G_0 \longrightarrow G$ be the inclusion map. Then ιf and $\mathrm{id}_G : G \longrightarrow G$ are both extensions of the maps $f_i : G_i \longrightarrow G$. Since the extension is unique, $\iota f = \mathrm{id}_G$, so ι is onto, hence $G = G_0$.

To see existence, let $\langle X_i \mid R_i \rangle$ be a presentation of G_i (via some mapping which will be suppressed), with $X_i \cap X_j = \emptyset$ for $i \neq j$. Let Y be a set of generators for A and, for each $y \in Y$, $i \in I$, let $a_{i,y}$ be a word in $(X_i^{\pm 1})^*$ representing $\alpha_i(y)$. Let

$$S = \left\{ a_{i,y} a_{j,y}^{-1} \mid y \in Y, \, i, j \in I, \, i \neq j \right\}.$$

Let G be the group with presentation $\langle \bigcup_{i \in I} X_i \mid \bigcup_{i \in I} R_i \cup S \rangle$. Then G is the desired free product with amalgamation. There is an obvious mapping $X_i \longrightarrow G$, which by Lemma 5.2 induces a homomorphism $f_i : G_i \longrightarrow G$. The existence and uniqueness of the mapping h in the definition follows by another application of Lemma 5.2. As an example, let G_1 be an infinite cyclic group generated by x and G_2 an infinite cyclic group generated by y. Let A be infinite cyclic, generated by a, and define α_1, α_2 by $\alpha_1(a) = x^2$, $\alpha_2(a) = y^3$. The presentation of the corresponding free product with amalgamation is $\langle x, y \mid x^2 = y^3 \rangle$, after some obvious modification. This group has various geometrical interpretations; see §1.5.2, Chap. I in [35].

If G is the free product of $\{G_i \mid i \in I\}$ with A amalgamated, we write $G = \underset{i \in I}{\bigstar}(G_i : A)$ (suppressing the α_i). If the family consists of two groups, say $\{G_1, G_2\}$, we write $G = G_1 *_A G_2$.

To see G has the desired properties, we have to investigate its structure. First, it is convenient to replace the maps α_i by an amalgam. Take the disjoint union $A \amalg \coprod_{i \in I} G_i$ and identify a and $\alpha_i(a)$, for all $a \in A$ and $i \in I$. Call the quotient set B; for notational convenience, we shall (inaccurately) identify A and each G_i with their isomorphic images in B, so that $B = \bigcup_{i \in I} G_i$ is an amalgam with $G_i \cap G_j = A$ for $i \neq j$. We consider non-empty words over B, that is, elements of B^+. We shall use commas and parentheses when writing such words, to avoid confusion with the product in the groups G_i. The maps f_i induce a mapping $f : B \longrightarrow G$, where $f(g) = f_i(g)$ for $g \in G_i$, and there is an extension to a semigroup homomorphism $\overline{f} : B^+ \longrightarrow G$ given by $\overline{f}(g_1, \ldots, g_n) = f(g_1) \ldots f(g_n)$. We call $\overline{f}(w)$ the element of G represented by w. Since $\bigcup_{i \in I} f_i(G_i)$ generates G and the f_i are homomorphisms, every element of G is represented by a word in B^+.

Definition. Let $w = (g_1, \ldots, g_n)$ be a word with $g_j \in G_{i_j}$, $1 \leq j \leq n$, $n \geq 1$. Then w is *reduced* if

(1) $i_j \neq i_{j+1}$ for $1 \leq j \leq n-1$
(2) $g_j \notin A$ for $1 \leq j \leq n$, unless $n = 1$.

If (1) fails, we can replace w by $(g_1, \ldots, g_j g_{j+1}, \ldots, g_n)$, representing the same element of G. Similarly, if (2) fails, we can replace w by a shorter word. Hence every element of G is represented by a reduced word (take a word of shortest length representing it).

We need to refine the idea of reduced word. In B, for each $i \in I$ choose a set T_i of representatives for the cosets $\{Ax \mid x \in G_i\}$, with $1 \in T_i$. Let $g \in G$ be represented by the reduced word $(g_1, \ldots, g_n)$ with $g_j \in G_{i_j}$. We can write:

$$g_n = a_n r_n \qquad\qquad (a_n \in A,\ r_n \in T_{i_n})$$
$$g_{n-1} a_n = a_{n-1} r_{n-1} \qquad\qquad (a_{n-1} \in A,\ r_{n-1} \in T_{i_{n-1}})$$
$$\vdots \qquad\qquad\qquad\qquad \vdots$$
$$g_1 a_2 = a_1 r_1 \qquad\qquad (a_1 \in A,\ r_1 \in T_{i_1}).$$

Thus $g_1 = a_1 r_1 a_2^{-1}$, $g_2 = a_2 r_2 a_3^{-1}, \ldots, g_n = a_n r_n$ and $(a_1, r_1, r_2, \ldots, r_n)$ is a word representing g.

Definition. A *normal* word is a word $(a, r_1, \ldots, r_n)$ where $a \in A$, $n \geq 0$, $r_j \in R_{i_j} \setminus \{1\}$ $(1 \leq i \leq n)$, where $i_j \in I$ and $i_j \neq i_{j+1}$ for $1 \leq j \leq n-1$.

From the discussion above, any element of G is represented by a normal word. If $a \in A$, it is represented by the normal word (a), and words of this form are the only normal words which are reduced. Of course, if $(a, r_1, \ldots, r_n)$ is a normal word with $n \geq 1$, then $(r_1, \ldots, r_n)$ is reduced.

Theorem 5.8. (Normal Form Theorem) *Any element of G is represented by a unique normal word.*

Proof. We have to show uniqueness. Let W be the set of normal words. We shall define an action of G on W, equivalently, a homomorphism $h : G \longrightarrow S(W)$, the symmetric group on W. By the defining property of G, it suffices to define, for $i \in I$, a homomorphism $h_i : G_i \longrightarrow S(W)$ such that $h_i \alpha_i = h_j \alpha_j$ for all $i, j \in I$. We continue to work in B, so we need homomorphisms h_i which agree on A.

For $i \in I$, let $W_i = \{(1, r_1, \ldots, r_n) \in W \mid r_1 \notin T_i\}$. We define a mapping $\theta_i : G_i \times W_i \longrightarrow W$ as follows. If $g \in G_i$, write $g = ar$, where $r \in T_i$, $a \in A$. Then

$$\theta_i(g, (1, r_1, \ldots, r_n)) = \begin{cases} (a, r, r_1, \ldots, r_n) & \text{if } r \neq 1 \text{ (i.e. } g \notin A) \\ (a, r_1, \ldots, r_n) & \text{if } r = 1 \end{cases}$$

It is easily seen that θ_i is bijective. Now G_i acts on $G_i \times W_i$ by left translation on the first coordinate, giving a homomorphism $\eta_i : G_i \longrightarrow S(G_i \times W_i)$, where $\eta_i(g)(x, w) = (gx, w)$. Set $h_i(g) = \theta_i \eta_i(g) \theta_i^{-1}$. Since η_i is a homomorphism, so is h_i. If $a \in A$, it is an easy exercise to see that

$$h_i(a)(a', r_1, \ldots, r_n) = (aa', r_1, \ldots, r_n)$$

and the right-hand side is independent of i. Thus the h_i define a homomorphism $h : G \longrightarrow S(W)$.

Let $g \in G$ be represented by the normal word $w = (a, r_1, \ldots, r_n)$. Then it is easy to see that $h(g)(1) = w$, so w is uniquely determined by g. $\qquad \square$

The proof is taken from Serre's tree notes [35], and is based on an argument of van der Waerden [40].

Corollary 5.9. (1) *The homomorphisms f_i are injective.*
(2) *No reduced word of length greater than 1 represents the identity element of G.*
(3) $f_i(G_i) \cap f_j(G_j) = f_i \alpha_i(A)$ *for $i \neq j$.*

Proof. (1) If $f_i(g) = 1$, write $g = ar$ with $a \in A$, $r \in T_i$. Then the normal word (a, r) (or (a), if $r = 1$) represents 1, hence by Theorem 5.8, $a = r = 1$, so $g = 1$.

(2) If a reduced word has length $n > 1$, the procedure above gives a normal word of length $n + 1$ representing the same element of G, which cannot be 1 by Theorem 5.8, as the normal word representing 1 is (1).

(3) If $i \neq j$, clearly $f_i \alpha_i(A) \subseteq f_i(G_i) \cap f_j(G_j)$. Suppose $g \in f_i(G_i) \cap f_j(G_j)$ and $g \notin f_i \alpha_i(A)$. Then g is represented by reduced words (a, r) and (a', r'), where $r \in T_i \setminus \{1\}$ and $r' \in T_j \setminus \{1\}$. By Theorem 5.8, $r = r'$, which is impossible as T_i, T_j intersect only in 1. $\qquad \square$

Conversely, (1) and (2) of the corollary imply the Normal Form Theorem. See [5, §1.4]. In view of Cor. 5.9(1), the f_i can be suppressed, to simplify the notation.

An important special case is when A is the trivial group. In this case, G is called the free product of the family $\{G_i \mid i \in I\}$, written $G = \ast_{i \in I} G_i$ (or $G = G_1 \ast G_2$ in the case of two groups). A reduced word is a word $(g_1, \ldots, g_n)$ such that $g_j \in G_{i_j}$,

$g_j \neq 1$ unless $n = 1$, and $i_j \neq i_{j+1}$ for $1 \leq j \leq n-1$. The Normal Form Theorem simplifies to: every element of G is represented by a unique reduced word. Also, the G_i embed in G. The universal mapping property simplifies to: given any collection of homomorphisms $h_i : G_i \longrightarrow H$, there is a unique extension to a homomorphism $h : G \longrightarrow H$. Further, the presentation above used to show existence of free products with amalgamation simplifies. Let $\langle X_i \mid R_i \rangle$ be a presentation of G_i (via some mapping which will be suppressed), with $X_i \cap X_j = \emptyset$ for $i \neq j$. Then $*_{i \in I} G_i$ has presentation $\langle \bigcup_{i \in I} X_i \mid \bigcup_{i \in I} R_i \rangle$. (This is obtained from the presentation above by taking the empty presentation of the trivial group, with no generators and no relations.)

As an example, $\langle x, y \mid x^2 = 1, \ y^2 = 1 \rangle$ is a presentation of the free product of two cyclic groups of order 2, and is called the infinite dihedral group. (It can be shown that its only proper, non-trivial quotients are the dihedral groups.) The free product of a cyclic group of order 2 and a cyclic group of order 3 has presentation $\langle x, y \mid x^2 = 1, \ y^3 = 1 \rangle$. This group is called the *modular group*, and is isomorphic to $\mathrm{PSL}_2(\mathbb{Z})$; see [26, Theorem 3.1].

HNN-extensions. Suppose B, C are subgroups of a group A and $\gamma : B \longrightarrow C$ is an isomorphism. In general, γ is not induced by an automorphism of A. However ([15]), it is always possible to embed A in a group G such that γ is induced by an inner automorphism of G. That is, there is an element $t \in G$ such that $tbt^{-1} = \gamma(b)$ for all $b \in B$.

To prove this, we consider the group G with presentation (via some mapping which is suppressed) $\langle \{t\} \cup X \mid R_1 \cup R_2 \rangle$, where

$$X = \{x_g \mid g \in A\}, \quad t \notin X$$
$$R_1 = \left\{ x_g x_h x_{gh}^{-1} \mid g, h \in A \right\}$$
$$R_2 = \left\{ t x_b t^{-1} x_{\gamma(b)}^{-1} \mid b \in B \right\}.$$

By Lemma 5.2 applied to the standard presentation of A, there is a homomorphism $f : A \to G$ given by $a \mapsto x_a$, and we have to show that f is injective.

Definition. The group G is called an HNN-extension with *base* A, *associated pair of subgroups* B, C and *stable letter* t.

This presentation is often abbreviated to $\langle t, A \mid \mathrm{rel}(A), tBt^{-1} = \gamma(A) \rangle$. Sometimes γ is suppressed and we write $\langle t, A \mid \mathrm{rel}(A), tBt^{-1} = C \rangle$, or even more extremely, just $\langle t, A \mid tBt^{-1} = C \rangle$.

More generally, Let $\langle Y \mid S \rangle$ be a presentation of A, let $\{b_j \mid j \in J\}$ be a set of words in $(Y^{\pm 1})^*$ representing a set of generators for B, and let c_j be a word representing $\gamma(b_j)$ (more accurately, γ(the generator represented by b_j)). Then

$$\langle t \cup Y \mid S \cup \{t b_j t^{-1} = c_j \mid j \in J\} \rangle$$

is also a presentation of G. This is left as an exercise, using Lemma 5.2.

As an example, $\langle x, y \mid xy^2 x^{-1} = y^3 \rangle$ is an HNN-extension, with base an infinite cyclic group A generated by y, stable letter x and associated pair the subgroups of

A generated by y^2, y^3 respectively. (The isomorphism γ is given by $\gamma(y^{2n}) = y^{3n}$ for $n \in \mathbb{Z}$.) This is a famous example of a non-Hopfian group, the *Baumslag-Solitar group*; see Theorem 4.9, Chap. IV in [25].

Returning to the general situation, since G is generated by $f(A) \cup \{t\}$, every element of G is represented in an obvious way by a word $(g_0, t^{e_1}, g_1, t^{e_2}, g_2, \ldots, t^{e_n}, g_n)$, where $n \geq 0$, $e_i = \pm 1$ and $g_i \in A$ for $0 \leq i \leq n$.

Definition. Such a word is *reduced* if it has no subword of the form t, b, t^{-1} with $b \in B$ or t^{-1}, c, t with $c \in C$.

Any element of G is represented by a reduced word (take a word of minimal length representing it).

Choose a set R_B of representatives for the cosets $\{Bg \mid g \in A\}$ and a set R_C of representatives for the cosets $\{Cg \mid g \in A\}$, with $1 \in T_B, T_C$.

Definition. A *normal word* is a reduced word $(g_0, t^{e_1}, r_1, t^{e_2}, r_2, \ldots, t^{e_n}, r_n)$, where $g_0 \in A$, $r_i \in R_B$ if $e_i = 1$ and $r_i \in R_C$ if $e_i = -1$.

Suppose $(g_0, t^{e_1}, g_1, t^{e_2}, g_2, \ldots, t^{e_n}, g_n)$ is a reduced word with $n \geq 1$. If $e_n = 1$, write $g_n = br$ with $b \in B$, $r \in R_B$. Then $(g_0, t^{e_1}, g_1, t^{e_2}, g_2, \ldots, g'_{n-1}, t^{e_n}, r)$, where $g'_{n-1} = g_{n-1}\gamma(b)$, represents the same element of G. If $e_n = -1$, write $g_n = cr$ with $c \in C$ and $r \in R_C$. Then $(g_0, t^{e_1}, g_1, t^{e_2}, g_2, \ldots, g'_{n-1}, t^{e_n}, r)$, where $g'_{n-1} = g_{n-1}\gamma^{-1}(c)$, represents the same element of G. Repetition of this procedure leads to a normal word representing the same element of G.

Theorem 5.10. (Normal Form Theorem) *Any element of G is represented by a unique normal word.*

Proof. This can again be proved using the van der Waerden method; we refer to [25, Chap. IV, Theorem 2.1] for the details.   □

Corollary 5.11. (1) *The homomorphism $f : A \longrightarrow G$ is injective.*
(2) (Britton's Lemma) *no reduced word $(g_0, t^{e_1}, g_1, t^{e_2}, g_2, \ldots, t^{e_n}, g_n)$ with $n > 0$ represents the identity element of G.*

Proof. This follows easily from the Normal Form Theorem and details are left to the reader.   □

As with free products with amalgamation, the Normal Form Theorem follows from the corollary. See [5, §1.5]. We also note that it is unnecessary to directly prove both Normal Form Theorems, for free products with amalgamation and for HNN-extensions, as one implies the other. See [5, Chap. 1, Exercises 23 and 24]. In view of Cor. 5.11(1), the mapping f can be suppressed.

More generally, given A and a family $\gamma_i : B_i \longrightarrow C_i$ $(i \in I)$ of isomorphisms, where B_i, C_i are subgroups of A, we can form the HNN-extension with presentation (in abbreviated form):

$$\langle t_i \ (i \in I), \ G \mid \mathrm{rel}(G), \ t_i B_i t_i^{-1} = \gamma_i(B_i) \ (i \in I) \rangle.$$

There are generalisations of the Normal Form Theorem and its corollary, but we shall not need these. For further properties of free products with amalgamation and HNN-extensions, and their uses, we refer to [5] and [25]. However, there is one result we shall need later. The proof is from [23]; for a different viewpoint, see [3] (just before Theorem 3.1).

Lemma 5.12. (1) *If $G = B *_A C$ and G, A are finitely generated, then B and C are finitely generated.*
(2) *if $G = \langle t, A \mid tBt^{-1} = C \rangle$ is an HNN-extension, and G, B are finitely generated, then A is finitely generated.*

Proof. (1) Assume G, A are finitely generated. It suffices by symmetry to show B is finitely generated. Suppose not. Since G is countable, so is B, so there are subgroups $A \lneqq B_1 \lneqq B_2 \lneqq \ldots$ of B with $B = \bigcup_{i=1}^{\infty} B_i$. Let G_i be the subgroup of G generated by $B_i \cup C$. (Note that G_i is isomorphic to $B_i *_A C$. For the inclusion mapping $B_i \to B$ and identity mapping $C \longrightarrow C$ have an extension to a homomorphism $h : B_i *_A C \longrightarrow B *_A C$. The image is G_i, and h is injective by Cor. 5.9.) Then $G = \bigcup_{i=1}^{\infty} G_i$ and $G_1 \lneqq G_2 \lneqq \ldots$ as $G_i \cap B = B_i$ by Cor. 5.9. This is a contradiction since G is finitely generated.

(2) Assume G and B are finitely generated (so C is, being isomorphic to B). Let D be the subgroup of G generated by $B \cup C$, so D is finitely generated. Suppose A is not finitely generated. Then there are subgroups $D \lneqq A_1 \lneqq A_2 \lneqq \ldots$ of A with $A = \bigcup_{i=1}^{\infty} A_i$. Let G_i be the subgroup of G generated by $A_i \cup \{t\}$. (By Cor. 5.11, G_i is isomorphic to an HNN-extension $\langle t, A_i \mid tBt^{-1} = C \rangle$.) Then $G = \bigcup_{i=1}^{\infty} G_i$ and $G_1 \lneqq G_2 \lneqq \ldots$ as $G_i \cap A = A_i$ by Cor. 5.11. This is a contradiction since G is finitely generated. $\square$

The Word Problem

Groups arising in geometry and topology are frequently given by presentations, and so it is desirable to be able to deduce information on a group from a presentation. One question is: given words u and v, do they represent the same element of the group, that is, does the relation $u = v$ hold? This holds if and only if $uv^{-1} = 1$ does, so it suffices to know whether or not a relation $w = 1$ holds. Informally, the word problem, formulated by Dehn, is to find a procedure with a finite set of instructions to decide, given $w \in X^{\pm 1}$, whether or not w represents 1 in G. He found such a procedure which works for a certain class of presentations, including the usual presentations of surface groups, now known as Dehn's algorithm.

This can easily be made precise. Let G be a group, X a set and $\varphi : X \longrightarrow G$ a mapping such that $\varphi(X)$ generates G. Put

$$W_\varphi(G) = \left\{ w \in (X^{\pm 1})^* \mid \overline{\varphi}(w) = 1_G \right\}.$$

(In Lemma 5.5, this is the set of words representing elements of $\mathrm{Ker}(f)$, so by Cor. 5.7, if $G = \langle X \mid R \rangle^\varphi$, it is determined by R.)

Definition. Assume X is finite. The word problem for G (relative to φ) is solvable if $W_\varphi(G)$ is a recursive language (the alphabet being $X^{\pm 1}$).

Example. The word problem for $F(X) = \langle X | \emptyset \rangle^i$, where X is a finite set and $i :$ $X \longrightarrow F(X)$ is the inclusion map, is solvable. In fact, $W_i(F(X))$ is deterministic. For define a PDA $M = (Q, F, A, \Gamma, \tau, q_0, z_0)$ recognising $W_i(F(X))$ by: $Q = \{q_0, q_1, q_2\}$, $F = \{q_0\}$, $A = X^{\pm 1}$, $\Gamma = A \cup \{z_0\}$ and τ consists of the transitions

$$(q_0, \varepsilon, z_0, q_1, z_0)$$
$$(q_1, y, z, q_2, yz) \quad (y, z \in A, \ y \neq z^{-1})$$
$$(q_1, y, y^{-1}, q_2, \varepsilon) \quad (y \in A)$$
$$(q_2, \varepsilon, z, q_1, z) \quad (z \in A)$$
$$(q_2, \varepsilon, z_0, q_0, z_0)$$

Thus M stores the reduced form of the word read from the tape in its stack, with z_0 at the bottom. When z_0 is read at the top of the stack, the reduced form is ε and M enters the final state q_0. The extra state q_2 is needed to make M deterministic. Note that M accepts ε, since there is a computation with just a single configuration, (q_0, ε, z_0).

We shall show that solvability of the word problem depends only on G, not on the choice of φ.

Definition. Let $\mathscr{L}$ be a class of languages, with possibly different finite alphabets. Then $\mathscr{L}$ is *closed under inverse homomorphism* if, given a monoid homomorphism $\varphi : A^* \to B^*$ (where A, B are finite sets) and a language $L \in \mathscr{L}$ with alphabet B, then $\varphi^{-1}(L) \in \mathscr{L}$.

Lemma 5.13. *Let $\mathscr{L}$ be a class of languages closed under inverse homomorphism. Let $\varphi : X \to G$, $\psi : Y \to G$ be maps, where G is a group, X and Y are finite, and $\varphi(X)$, $\psi(Y)$ generate G. Then $W_\varphi(G) \in \mathscr{L}$ if and only if $W_\psi(G) \in \mathscr{L}$.*

Proof. It suffices by symmetry to show that if $W_\psi(G) \in \mathscr{L}$, then $W_\varphi(G) \in \mathscr{L}$. Let $A = X^{\pm 1}$, $B = Y^{\pm 1}$. Extend φ, ψ to A^*, B^* respectively ($\overline{\varphi}(x^{-1}) = \varphi(x)^{-1}$ for $x \in X$, and $\overline{\varphi}(a_1 \dots a_n) = \overline{\varphi}(a_1) \dots \overline{\varphi}(a_n)$ for $a_i \in A$, etc.). Note that $\overline{\varphi}$, $\overline{\psi}$ are surjective monoid homomorphisms.

We claim there is a monoid homomorphism f making the diagram commutative. For $x \in X$, choose $w \in B^*$ such that $\overline{\psi}(w) = \varphi(x)$, then put $f(x) = w$, $f(x^{-1}) = w^{-1}$. Then extend to A^* by defining $f(a_1 \dots a_n) = f(a_1) \dots f(a_n)$, for $a_i \in A$.

$$\begin{array}{ccc} & & A^* \\ & {\scriptstyle f} \downarrow & \searrow {\scriptstyle \overline{\varphi}} \\ & B^* & \xrightarrow[\overline{\psi}]{} \ G \end{array}$$

Now $W_\varphi(G) = f^{-1}(W_\psi(G))$, so $W_\psi(G) \in \mathscr{L}$ implies $W_\varphi(G) \in \mathscr{L}$. $\square$

Thus if $\mathscr{L}$ is closed under inverse homomorphism, it makes sense to say a group G has word problem in $\mathscr{L}$. The following classes are closed under inverse homomorphism (references are to [21]).

(1)	regular	Theorem 3.5
(2)	deterministic	Theorem 10.4
(3)	context-free	Theorem 6.3
(4)	context-sensitive	Exercise 9.10 and solution
(5)	recursive	§11.1
(6)	recursively enumerable	§11.1.

Thus we can speak of a group G having regular or context-free word problem, etc. Another useful fact is the following lemma.

Lemma 5.14. *Let $\mathcal{L}$ be a class of languages such that:*

(1) *$\mathcal{L}$ is closed under inverse homomorphism;*
(2) *If $L \in \mathcal{L}$ and R is a regular language, then $L \cap R \in \mathcal{L}$.*

If G is a finitely generated group with word problem in $\mathcal{L}$, and H is a finitely generated subgroup of G, then H has word problem in $\mathcal{L}$.

Proof. Let $\varphi : X \longrightarrow G$ be a mapping such that X is finite and $\varphi(X)$ generates G, and let $\psi : Y \longrightarrow H$ be such that Y is finite and $\psi(Y)$ generates H. We assume $X \cap Y = \emptyset$. Define $\theta : X \cup Y \longrightarrow G$ by $\theta|_X = \varphi$, $\theta|_Y = \psi$, so $\theta(X \cup Y)$ generates G. By Lemma 5.13, $W_\theta(G) \in \mathcal{L}$, and $W_\psi(H) = W_\theta(G) \cap (Y^{\pm 1})^*$. By Lemma 1.5, $(Y^{\pm 1})^*$ is regular, so by assumption (2), $W_\psi(H) \in \mathcal{L}$.   □

All the language classes listed after Lemma 5.13 satisfy the hypotheses of Lemma 5.14. With the exception of deterministic and context-free, these classes are closed under intersection and contain the class of regular languages. (See Lemma 2.2, Lemma 3.3, Lemma 1.5 and [20, Theorem 9.6].) For deterministic and context-free languages, see Lemma 4.16.

We shall need a generalisation of the idea of a class closed under inverse homomorphism. This involves the notion of *generalised sequential machine*, abbreviated to gsm. This is an elaboration of a FSA which produces output. In fact, producing output is their only function, and they are not intended for language recognition.

Definition. A generalised sequential machine is a sextuple $S = (Q, F, A, B, \tau, q_0)$, where Q, A and B are finite sets (the set of states, the input alphabet and the output alphabet respectively), $F \subseteq Q$ (the set of final states), $q_0 \in Q$ (the initial state) and τ is a finite subset of $Q \times A \times B^* \times Q$ (the set of transitions).

A *computation* of S is a sequence $q_0, (a_1, u_1), q_1, (a_2, u_2), \ldots, (a_n, u_n), q_n$, where $n \geq 0$, $q_i \in Q$ $(0 \leq i \leq n)$, $u_i \in B^*$ $(1 \leq i \leq n)$ and $(q_{i-1}, a_i, u_i, q_i) \in \tau$ for $1 \leq i \leq n$. The computation is *successful* if $q_n \in F$. The *input* of the computation is $a_1 \ldots a_n \in A^*$ and the *output* is $u_1 \ldots u_n \in B^*$.

As with a FSA, we can form the *transition diagram* of S. This is a directed graph with vertex set Q and an edge from q to q' for each transition (q, a, u, q'), with label (a, u)[1]. Then paths in the transition diagram starting at q_0 are in 1-1 correspondence with computations of S. Both the input and output can be read off from the labels on the edges of the path. For $w \in A^*$, we define

---

[1] The label is often denoted by $a|u$ in the literature.

$$f_S(w) = \{u \in B^* \mid \text{there is a successful computation with input } w, \text{ output } u\}$$

and for $u \in B^*$,

$$f_S^{-1}(u) = \{w \in A^* \mid \text{there is a successful computation with input } w, \text{ output } u\}.$$

Thus f_S is a mapping from A^* to the set of subsets of B^*. Note that f_S^{-1} is not necessarily the inverse of f_S, in the usual sense.

If L is a language with alphabet A, we define $f_S(L) = \bigcup_{w \in L} f_S(w)$, and if L' is a language with alphabet B, put $f_S^{-1}(L') = \bigcup_{u \in L'} f_S^{-1}(u)$. We call f_S a gsm mapping and f_S^{-1} an inverse gsm mapping.

The gsm S is called *deterministic* if, for $q \in Q$ and $a \in A$, there is at most one transition starting with q, a. Then $f_S(w)$ is either empty or contains a single element, so f_S may be viewed as a partial function from A^* to B^*, and f_S^{-1} is then the inverse of f_S.

Definition. A class $\mathscr{L}$ of languages is *closed under inverse gsm mappings* if whenever $L \in \mathscr{L}$ has alphabet B and S is a gsm with output alphabet B, then $f_S^{-1}(L) \in \mathscr{L}$.

We can also define what is meant by a class closed under inverse deterministic gsm mappings (restrict S in the definition to be deterministic). The class of deterministic languages is closed under inverse deterministic gsm mappings. This will be used later, so a proof is given at the end of Appendix A. The other classes listed after Lemma 5.13 are closed under inverse gsm mappings. See [21, Theorem 11.2].

Suppose $f : A^* \longrightarrow B^*$ is a monoid homomorphism. Construct a gsm

$$S = (\{q_0\}, \{q_0\}, A, B, \tau, q_0)$$

where τ consists of the transitions $(q_0, a, f(a), q_0)$ for $a \in A$. Then S is deterministic, f_S is total and $f_S = f$. It follows that, if $\mathscr{L}$ is closed under inverse deterministic gsm mappings, it is closed under inverse homomorphism, so Lemma 5.13 applies to $\mathscr{L}$. The argument of the next lemma is part of the proof of Lemma 5 in [17].

Lemma 5.15. *Let $\mathscr{L}$ be a class of languages closed under inverse deterministic gsm mappings. Let G be a finitely generated group, and let H be a subgroup of finite index. If H has word problem in $\mathscr{L}$, then so does G.*

Proof. Let T be a transversal for $\{Hg \mid g \in G\}$ and let $\varphi : X \to H$ be a mapping with X finite such that $\varphi(X)$ generates H. We can assume $1 \in T$ and $X \cap T = \emptyset$. Let $Y = X \cup T$ and define $\psi : Y \to G$ by: $\psi|_X = \varphi$, $\psi(t) = t$ for $t \in T$. Then $\psi(Y)$ generates G. There is a gsm $S = (T, \{1\}, A, B, \tau, 1)$, where $A = Y^{\pm 1}$, $B = X^{\pm 1}$ and τ is defined as follows.

For each $y \in A$ and $t \in T$, choose $h_{t,y} \in B^*$ such that $ty = h_{t,y}t'$ holds in G, where $t' \in T$. Then τ contains the transition $(t, y, h_{t,y}, t')$. Clearly S is deterministic. If a computation has input w, output u and ends in state t, then $w = ut$ holds in G via ψ. It follows that $W_\psi(G) = f_S^{-1}(W_\varphi(H))$. $\square$

Definition. A group is *finitely presented* if it has a presentation $\langle X \mid R \rangle$ with X, R finite. It is *recursively presented* if it has a presentation $\langle X \mid R \rangle$ with X finite and R recursively enumerable.

Lemma 5.16. *Suppose $G = \langle X \mid R \rangle^{\varphi}$ with X finite, R recursively enumerable. Then*

(1) $W_{\varphi}(G)$ *is recursively enumerable*
(2) G *has a presentation $G = \langle Y \mid S \rangle^{\psi}$ with Y finite, S recursive.*

Proof. (1) If $X = \{x_1, \ldots, x_n\}$, number the elements of $X^{\pm 1}$ as

$$\{x_1, \ldots, x_n, x_1^{-1}, \ldots, x_n^{-1}\} = \{y_1, \ldots, y_{2n}\}$$

and let θ be the Gödel numbering $\varphi_2 : (X^{\pm 1})^* \longrightarrow \mathbb{N}$ defined after Lemma 3.8. We leave as exercises the following facts.

(i) There is a primitive recursive function $\mathrm{pr} : \mathbb{N}^2 \longrightarrow \mathbb{N}$ such that $\mathrm{pr}(\theta(u), \theta(v)) = \theta(uv)$.
(ii) There is a primitive recursive function $\mathrm{inv} : \mathbb{N} \longrightarrow \mathbb{N}$ such that $\mathrm{inv}(\theta(u)) = \theta(u^{-1})$.
(iii) $R^{\pm 1}$ is r.e.

Now let $C = \{uru^{-1} \mid u \in (X^{\pm 1})^*, \ r \in R^{\pm 1}\}$. Then C is r.e. For there are recursive functions $g : \mathbb{N} \longrightarrow \mathbb{N}$ with $g(\mathbb{N}) = \theta((X^{\pm 1})^*)$ and $h : \mathbb{N} \longrightarrow \mathbb{N}$ with $h(\mathbb{N}) = \theta(R^{\pm 1})$. Define $f : \mathbb{N}^2 \longrightarrow \mathbb{N}$ by $f(x, y) = \mathrm{pr}(\mathrm{pr}(g(x), h(y)), \mathrm{inv}(g(x)))$, then let $\bar{f} = f \circ J^{-1}$, where J is the function in Chapter 2, Exercise 3. Then $\bar{f}$ is recursive and $\bar{f}(\mathbb{N}) = \theta(C)$.

It follows from Lemma 3.14 that C^* is r.e. Let f^* be a recursive function with $\theta(C^*) = f^*(\mathbb{N})$. Now if w is a word, then $w \in W_{\varphi}(G)$ if and only if w represents the same element of $F(X)$ as some element $u \in C^*$, by Cor. 5.7. Equivalently, $w^{-1}u \in W$, where $W = W_i(F(X))$, for some $u \in C^*$. Also, W is recursive (indeed deterministic), so the characteristic function χ of $\theta(W)$ is recursive. Thus

$$w \in W_{\varphi}(G) \Longleftrightarrow \chi(\theta(w^{-1}u)) = 1$$

for some $u \in C^*$. If we define $k : \mathbb{N}^2 \longrightarrow N$ by

$$k(m, n) = \begin{cases} m & \text{if } m \in \theta((X^{\pm 1})^*) \wedge \chi(\mathrm{pr}(\mathrm{inv}(m), f^*(n))) = 1 \\ 1 & \text{otherwise} \end{cases}$$

then k is recursive, hence so is $\bar{k} = k \circ J^{-1}$, and $\bar{k}(\mathbb{N}) = \theta(W_{\varphi}(G))$. (Note that $1 = \theta(\varepsilon)$.)

(2) The proof is known as "Craig's trick". Take a letter $y \notin X^{\pm 1}$ and as alphabet take $A = X^{\pm 1} \cup \{y\}$ (we can assume $R \neq \emptyset$). Number elements of A so that y has the highest number, say z (so $z = 2|X| + 1$). Take as Gödel numbering $\theta : A^* \to \mathbb{N}$ the numbering φ_2 defined after Lemma 3.8. Then there is a recursive function $f : \mathbb{N} \to \mathbb{N}$ such that $f(\mathbb{N}) = \theta(R)$. Let $w_i = \theta^{-1}(f(i))$, so $R = \{w_i \mid i \in \mathbb{N}\}$. Put $Y = X \cup \{y\}$

and $S = \{y\} \cup \{w_i y^i \mid i \in \mathbb{N}\}$. We claim that S is recursive and $G = \langle Y \mid S \rangle^\psi$, where $\psi|x = \varphi$, $\psi(y) = 1$.

To see S is recursive, we first note that $U = \{wy^i \mid w \in (X^{\pm 1})^*, \, i \in \mathbb{N}\}$ is recursive. Explicitly, $n \in \theta(U) \Leftrightarrow \exists l \leq \log_2(n)\big(P(l,n) \wedge Q(l,n)\big) \wedge n \in \theta(A^*)$, where

$$P(l,n) \Leftrightarrow \forall i \leq l(i > 0 \Rightarrow \log_{p_i}(n) < z) \text{ and}$$
$$Q(l,n) \Leftrightarrow \forall i \leq \log_2(n)(i > l \Rightarrow \log_{p_i}(n) = z).$$

Further, if $n = \theta(wy^i)$, where $wy^i \in U$, then putting $c(n) = \mu i \leq \log_2(n)(\log_{p_i}(n) = z) \dot{-} 1$ and $h(n) = \log_2(n) \dot{-} c(n)$, it follows that $i = h(n)$. Also, $\theta(w) = g(n)$, where $g(n) = 2^{c(n)} \prod_{j=1}^{c(n)} p_i^{\log_{p_i}(n)}$. Thus

$$n \in \theta(S) \Leftrightarrow n \in \theta(U) \wedge f(h(n)) = g(n).$$

Since h and g are primitive recursive, this shows S is recursive. The proof that $G = \langle Y \mid S \rangle^\psi$ is left to the reader. (Let $H = \langle Y \mid S \rangle^\psi$. Use Lemma 5.2 to define homomorphisms $G \longrightarrow H$ and $H \longrightarrow G$; then use Lemma 5.2 again to show these are inverse isomorphisms.)   □

We now mention, without proof, the two major results concerning solvability of the word problem.

Boone-Novikov Theorem. *There is a finitely presented group with unsolvable word problem.*

Higman Embedding Theorem. *A finitely generated group is recursively presented if and only if it can be embedded in a finitely presented group.*

For proofs, see [5, Chap. 9], [25, Chap. 4] or [32, Chap. 12]. Although these give different proofs of the Boone-Novikov Theorem, in all cases the proof of the Higman Embedding Theorem depends on a construction used in the Boone-Novikov Theorem. On the other hand, it is easy to deduce the Boone-Novikov Theorem from Higman's Embedding Theorem, as we now show.

First, it is not difficult to exhibit a recursively presented group with unsolvable word problem. Let S be a r.e., non-recursive subset of $\mathbb{N}$ (see Prop. 3.6). Let F be a free group with basis $\{a,b\}$; then $\{a^i b a^{-i} \mid i \in \mathbb{N}\}$ is a basis for a free subgroup of F (see the exercises at the end of the chapter). Hence $\{a^i b a^{-i} \mid i \in S\}$ is a basis for the subgroup G of F it generates, and $a^i b a^{-i} \in G$ if and only if $i \in S$ (for example, using the criterion in Exercise 3(c)).

Let G have presentation

$$\langle a,b,c,d \mid a^i b a^{-i} = c^i d c^{-i} \, (i \in S) \rangle$$

via φ, say. It follows that G is the free product of two free groups of rank 2, amalgamating two free subgroups of countably infinite rank (see Exercise 4 for the definition of rank). Then $a^i b a^{-i} c^i d^{-1} c^{-i} \in W_\varphi(G)$ if and only if $i \in S$, by Cor. 5.9. Let $\theta : (\{a,b,c,d\}^{\pm 1})^* \longrightarrow \mathbb{N}$ be the Gödel numbering φ_2 defined after Lemma 3.8.

It is left to the reader to show there is a recursive function $f : \mathbb{N} \longrightarrow \mathbb{N}$ such that $f(i) = \theta(a^i b a^{-i} c^i d^{-1} c^{-i})$ for all i. Then $f(S)$ is r.e. by Lemma 3.3, hence G is recursively presented. Also, $i \in S$ if and only if $f(i) \in \theta(W_\varphi(G))$, hence $W_\varphi(G)$ is not recursive, otherwise S would be recursive.

Now to deduce the Boone-Novikov Theorem from the Higman Embedding Theorem, let G be a recursively presented group with unsolvable word problem. By Higman's Theorem, G embeds in a finitely presented group. This group has unsolvable word problem by Lemma 5.14.

The word problem is solvable if $W_\varphi(G)$ is recursive. One can also ask what happens if $W_\varphi(G)$ belongs to one of the other classes in the hierarchy at the end of Chapter 4. This has a nice answer in the case of regular and context-free languages.

Theorem 5.17 (Anisimov). *A finitely generated group has regular word problem if and only if it is finite.*

Proof. Assume G is finite. Then the standard presentation of G

$$\langle x_g \mid x_g x_h = x_{gh} \ (g, \ h \in G) \rangle^\varphi$$

where $\varphi(x_g) = g$, is finite.

Construct a FSA $M = (Q, F, A, \tau, q_0)$ by putting

$$Q = \{x_g \mid g \in G\}, \ A = \{x_g^{\pm 1} \mid g \in G\}, \ q_0 = x_1, \ F = \{q_0\}$$

and letting τ consist of the transitions

$$(x_g, x_h, x_{gh}) \quad (g, \ h \in G)$$
$$(x_g, x_h^{-1}, x_{gh^{-1}}) \quad (g, \ h \in G).$$

Then M recognises $W_\varphi(G)$.

Conversely, assume G is infinite, and $\varphi : X \to G$ is a mapping such that $\varphi(X)$ generates G, where X is finite. Given a natural number n, there is an element $g \in G$ such that $\overline{\varphi}(u) \neq g$ for any word u with $|u| \leq n$. (There are only finitely many words in $X^{\pm 1}$ of length at most n.) Let $w \in X^{\pm 1}$ be of minimal length such that $\overline{\varphi}(w) = g$. Then $|w| > n$ and for all subwords $u \neq \varepsilon$ of w, $\overline{\varphi}(u) \neq 1$ (otherwise w could be shortened by deleting a subword, without changing $\overline{\varphi}(w)$).

Let M be a deterministic FSA with tape alphabet $X^{\pm 1}$. Let n be the number of states of M, and choose a word w with $|w| > n$ and such that, for all subwords $u \neq \varepsilon$ of w, $\overline{\varphi}(u) \neq 1$. Starting M with w on the tape, there are prefixes w_1 and $w_1 w_2$ of w, with $w_2 \neq \varepsilon$, such that M, after reading w_1 and $w_1 w_2$, is in the same state. Then either M accepts both $w_1 w_1^{-1}$ and $w_1 w_2 w_1^{-1}$, or it rejects them both. (It will be in the same state after reading both.) Since $\overline{\varphi}(w_1 w_1^{-1}) = 1$, but $\overline{\varphi}(w_1 w_2 w_1^{-1}) \neq 1$ (because $\overline{\varphi}(w_2) \neq 1$ by choice of w), M cannot recognise $W_\varphi(G)$. $\qquad \square$

The corresponding result for context-free languages is that a group has context-free word problem if and only if it has a free subgroup of finite index. We begin by showing this under an additional assumption, following Muller and Schupp [27].

This depends on a characterisation of context-free groups in terms of what are called Cayley graphs, so we begin by describing these.

Cayley Graphs. Again let $\varphi : X \to G$ be a mapping such that $\varphi(X)$ generates the group G. Put $A = X^{\pm 1}$. We form a directed graph as follows. The set of vertices is G and the set of edges is $G \times X$. The edge $e = (g, x)$ starts at g and ends at $g\varphi(x)$, and is given label x. For every such edge we add an opposite edge $\bar{e}$, from $g\varphi(x)$ to g, with label x^{-1}. We also define $\bar{\bar{e}}$ to be e, giving an involution without fixed points on the set of edges. The resulting graph is the *Cayley graph* of G with respect to φ, denoted by $\Gamma(G, \varphi)$. The labelling of edges extends to a labelling of paths, where labels are in A^*. If p is a path, viewed as a sequence of edges $e_1, \ldots, e_n$, the *opposite path* $\bar{p}$ is the path $\bar{e}_n, \ldots, \bar{e}_1$, and the *length* of p is n. If w is the label on p, the label on $\bar{p}$ is w^{-1}. If g is a vertex and $y \in X^{\pm 1}$, there is a unique edge starting at g with label y, and a unique edge ending at g with label y. Consequently, given $w \in A^*$, and $g \in G$, there is a unique path starting at g with label w, and it ends at $g\overline{\varphi}(w)$. Hence, if w is the label on a path, then $\overline{\varphi}(w) = 1$ if and only if the path is closed. (Note that, at each vertex v, there is a *trivial path*, of length 0, from v to v. See [5, §5.1] for further discussion. We give trivial paths label ε.) Since $\varphi(X)$ generates G, it follows that $\Gamma(G, \varphi)$ is connected (i.e. there is a path joining any two vertices).

Suppose $a, b \in G$, and we take paths in $\Gamma(G, \varphi)$ from 1 to a, from a to b, and from b to 1, with labels u, v, w, respectively. Also, take a second path from a to b with label z, giving the situation illustrated in the left-hand picture below.

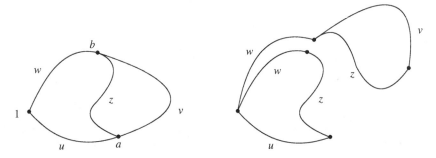

Figure 5.2

(The paths have been drawn in the plane for clarity-in the Cayley graph, they can have more intersections than illustrated, and need not be simple closed curves.)

There are closed paths p_1 at 1 with label uvw and p_2 with label uzw, and a closed path p_3 at b with label $z^{-1}v$. We obtain a closed path p_3' at 1 by traversing the path from b to 1 in the opposite direction, then p_3, then the path from b to 1, so p_3' has label $w^{-1}z^{-1}vw$. Note that $uvw =_{F(X)} (uzw)w^{-1}(z^{-1}v)w$. This can be visualised by "unstitching" the paths p_2 and p_3', so they meet at a single point, as in the right-hand diagram. The left-hand side uvw is the label reading anticlockwise around the boundary of the left-hand figure (i.e. p_1), and the right-hand side is the label reading anticlockwise around the boundary of the right-hand figure. We can continue to unstitch the two closed paths p_2 and p_3 in a similar manner, choosing paths joining two vertices they pass through to play the rôle of the path from a to b with label z in the first unstitching. This can be iterated, and always gives a product of conjugates of relators which hold in G (the labels on p_2, p_3 etc), equal in $F(X)$ to uvw. In some circumstances, by suitable choice of paths to unstitch, this can lead to a finite presentation of G. Examples are given in Lemma 5.24 and Theorem 5.29.

The group G acts on the vertices by left translation, and similarly on edges by $h(g,x) = (hg,x)$, $h\overline{(g,x)} = \overline{(hg,x)}$, so G acts on $\Gamma(G,\varphi)$ as graph automorphisms, transitively on the vertices, and preserving opposite edges: $g\bar{e} = \overline{ge}$ for $g \in G$ and edges e.

For $g, h \in G$, define

$$d(g,h) = \text{the length of a shortest path from } g \text{ to } h.$$

Then $d : G \times G \to \mathbb{N}$ is a metric, called the *path metric* on $\Gamma(G,\varphi)$. Since G acts as graph automorphisms, d is G-invariant, that is, $d(ag,ah) = d(g,h)$ for $a \in G$.

There is one final observation about Cayley graphs. As with presentations, the mapping φ is often suppressed in practice, and $\Gamma(G,\varphi)$ is written $\Gamma(G,X)$.

Now to characterise context-free groups by means of Cayley graphs, some discussion of plane polygons, and their triangulations, is needed. By a plane polygon P we mean a finite succession of arcs in the plane, joined together at their endpoints (the *vertices*), which form a simple closed curve, together with the interior of the curve. The interior is not required to be convex. The arcs are called boundary edges of P.

A *triangulation* of a plane polygon P is a decomposition of P into triangles, so that if two different triangles meet, they meet in an edge or a vertex. The edges of the triangles may be arcs rather than straight line segments. The original edges and vertices in the boundary of P must be edges and vertices of the triangles. A *diagonal triangulation* of P is a triangulation of P which has no vertices except the original vertices on the boundary of P. However, we allow "1-gons" (loops, i.e. simple closed curves with a point nominated as the vertex) and "bigons" (two arcs with the same endpoints) as polygons, and view these as triangulated.

We call a triangle in a diagonal triangulation of P *critical* if it has two edges which are boundary edges of P. Also, ∂P denotes the sequence of boundary edges of P reading anticlockwise around P. It is defined only up to cyclic permutation, as a starting vertex has not been specified.

Lemma 5.18. *Let T be a diagonal triangulation of a plane polygon P with at least two triangles. Then there are at least two critical triangles.*

Proof. The proof is by induction on the number of triangles. If there are two triangles, or all triangles having an edge on P are critical, the result is clear. Otherwise, choose a triangle T with exactly one edge e in the boundary of P. Starting at the vertex of T not incident with e, we can write $\partial P = w_1 e w_2$ and $\partial T = e_2 e e_1$. The triangulation of P induces diagonal triangulations of the polygon P_1 bounded by $w_1 e e_1$ and the polygon P_2 bounded by $w_2 e_2 e$, both having at least two triangles.

This is illustrated by the pictures below, using curved lines to represent w_1 and w_2.

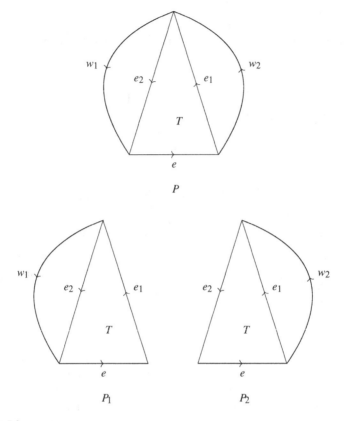

Figure 5.3

Thus the triangulations of P_1 and P_2 have at least two critical triangles by induction, so have at least one critical triangle other than T. Hence the triangulation of P has at least two critical triangles. □

Lemma 5.19. *Let P be a plane polygon with at least three boundary edges, having a diagonal triangulation. If the boundary edges of P are divided into three consecutive arcs, each with at least one edge, then some triangle has vertices on all three arcs.*

Proof. Let P be a polygon with a diagonal triangulation, and colour the vertices of P with three colours: red, white and blue. View red and white, white and blue, and blue and red as consecutive pairs of colours. We claim that, if consecutive vertices (reading round the boundary) have the same or consecutive colours and all three colours are used, then there is a triangle having vertices of all three colours. The proof is by induction on the number of triangles, and the result clearly holds when this is 1. Otherwise, there is a critical triangle T, say, by Lemma 5.18. If T has vertices of all three colours then T is the desired triangle. Otherwise, let P' be the polygon obtained by removing the two edges of T in the boundary of P. It is easily seen that P' has vertices of all three colours and consecutive vertices have the same or consecutive colours and the result follows by induction.

To prove the lemma, suppose $\partial P = \alpha\beta\gamma$, where α etc. are the sequences of edges of the three consecutive arcs, reading anticlockwise around P. Colour all vertices of α except the last red, all vertices of β except the last white and all vertices of γ except the last blue. Then a triangle with vertices of different colours is the desired triangle. $\square$

The characterisation of context-free groups in terms of Cayley graphs uses the following idea. Let $\varphi : X \longrightarrow G$ be a mapping such that $\varphi(X)$ generates G. Let α be a non-trivial closed path in $\Gamma(G, \varphi)$, with label $w = y_1 \ldots y_n$, so $\overline{\varphi}(w) = 1$. Let P be a plane polygon with n boundary edges. Write w anticlockwise around the boundary of P (with each edge labelled by a letter of w).

Definition. Let K be a positive real number. A K-triangulation of α is a diagonal triangulation of such a polygon P with a label in $(X^{\pm 1})^*$ assigned to each new edge such that

(1) reading around the boundary of each triangle gives a word u with $\overline{\varphi}(u) = 1$
(2) if u is the label on an edge of the triangulation then $|u| \leq K$.

Before stating the characterisation of context-free groups, a lemma is needed. We call a context-free grammar *reduced* if it has no useless symbols (see the definition before Lemma 4.5).

Lemma 5.20. *Let $\varphi : X \longrightarrow G$ be a mapping such that $\varphi(X)$ generates G and X is finite. Let E be a reduced context-free grammar with $W_\varphi(G) = L_E$. If A is a variable of E and $A \overset{\bullet}{\longrightarrow} u$, $A \overset{\bullet}{\longrightarrow} v$, where u, v are terminal strings, then $\overline{\varphi}(u) = \overline{\varphi}(v)$.*

Proof. Since E is reduced, there exist α, β such that $S \overset{\bullet}{\longrightarrow} \alpha A \beta$, and terminal strings w_1, w_2 such that $\alpha \overset{\bullet}{\longrightarrow} w_1$ and $\beta \overset{\bullet}{\longrightarrow} w_2$, hence $S \overset{\bullet}{\longrightarrow} w_1 u w_2$ and $S \overset{\bullet}{\longrightarrow} w_1 v w_2$, so $w_1 u w_2$, $w_1 v w_2 \in W_\varphi(G)$. Thus $\overline{\varphi}(w_1 u w_2) = \overline{\varphi}(w_1 v w_2) = 1$, and since $\overline{\varphi}$ is a monoid homomorphism, $\overline{\varphi}(u) = \overline{\varphi}(v)$. $\square$

Theorem 5.21. *Let $\varphi : X \longrightarrow G$ be a mapping such that $\varphi(X)$ generates G and with X finite. Then G has context-free word problem if and only if there is a constant K such that every non-trivial closed path in $\Gamma(G, \varphi)$ can be K-triangulated.*

Proof. Suppose G is context-free, so by Cor. 1.2, $W_\varphi(G) \setminus \{1\}$ is context-free. (During this proof 1 will denote the empty string ε.) By Theorem 4.7 there is a grammar E in Chomsky normal form such that $L_E = W_\varphi(G) \setminus \{1\}$. We have to find a constant K with the property claimed in the theorem. If $L_E = \emptyset$, then $X = \emptyset$ (otherwise $xx^{-1} \in L_E$ for any $x \in X$). Thus $\Gamma(G, \varphi)$ has a single vertex and no edges, so has no non-trivial paths. We can therefore assume $L_E \neq \emptyset$. Using the procedure of Lemma 4.5 gives a grammar still in Chomsky normal form, so we can assume E is reduced. If A is a variable of E, it is generating (as noted before Lemma 4.5), so we can choose a string of terminals u_A such that $A \overset{\bullet}{\longrightarrow} u_A$. (To minimise the value of K, we take u_A of shortest possible length. Note that $u_A \neq 1$ as E is in Chomsky normal form, so application of a production to a word does not decrease its length.) Let α be a non-trivial closed path in $\Gamma(G,\varphi)$, so its label $w = y_1 \ldots y_n \in W_\varphi(G)$, and write w round the boundary of an n-gon P in the plane. If $n \leq 2$ then by convention P is triangulated, and if $n = 3$, P is a triangle. In these cases, the labels on edges have length 1.

Assume that $n \geq 4$. Take a derivation of w from S (the start symbol). It must have the form $S, AB, \ldots, w_1 w_2 = w$, where $A \xrightarrow{\bullet} w_1$ and $B \xrightarrow{\bullet} w_2$. This divides the boundary of P into two arcs with labels w_1 and w_2. Suppose w_1 and w_2 both have length at least two. Construct an edge with label u_B from the vertex at which the arc with label w_1 ends to the vertex at which it begins, with label u_B.

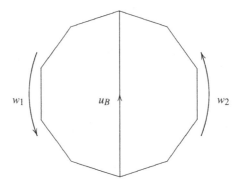

Figure 5.4

By Lemma 5.20, $\overline{\varphi}(u_B) = \overline{\varphi}(w_2)$, and since $\overline{\varphi}(w_1 w_2) = \overline{\varphi}(w) = 1$, $\overline{\varphi}(u_B) = \overline{\varphi}(w_1^{-1})$. We now have two polygons, with boundary labels $w_1 u_B$ and $w_2 u_B^{-1}$, with fewer sides. These labels represent 1 in G, so are labels on closed paths in $\Gamma(G,\varphi)$. If one of w_1, w_2 has length 1, say $w_1 = a$ where a is a terminal, then $S \xrightarrow{\bullet} aw_2$ where $|w_2| \geq 3$. The derivation of w from S, assuming it is leftmost, will then have the form $S, AB, aB, aCD, \ldots, aw_3 w_4$, where $w_2 = w_3 w_4$ and $C \xrightarrow{\bullet} w_3$, $D \xrightarrow{\bullet} w_4$, and at least one of w_3, w_4 has length at least 2, say w_4. This divides the boundary of P into an edge with label a and two arcs with labels w_3 and w_4. Draw an edge with label u_D from the vertex at which the arc with label w_4 begins to the vertex at which it ends. Once again this gives two polygons with fewer sides and with boundary labels $aw_3 u_D$ and $w_4 u_D^{-1}$. Since $\overline{\varphi}(u_D) = \overline{\varphi}(w_4)$ by Lemma 5.20 and $\varphi(aw_3 w_4) = \varphi(w) = 1$, these are labels on closed paths in $\Gamma(G,\varphi)$. (It is left to the reader to draw a picture for this case, and to deal with the cases not considered.)

Iteration of this procedure on the smaller polygons, treating labels of the form u_A just like terminals, eventually gives a diagonal triangulation of P with all labels on new edges of the form u_A, where A is a variable of E. Let $K = \max_{A \in V_N} |u_A|$, where V_N is the set of variables of E, Then $K \geq 1$, so we have constructed a K-triangulation of α, hence every non-trivial closed path can be K-triangulated.

Conversely, suppose there exists K such that every non-trivial closed path in $\Gamma(G,\varphi)$ can be K-triangulated. We construct a context-free grammar E with $L_E =$

$W_\varphi(G) \setminus \{1\}$. The set of terminals is $X^{\pm 1}$. For $u \in (X^{\pm 1})^*$ with $|u| \leq K$, there is a corresponding variable A_u. For each relation $u = vw$ which holds in G via φ, with $|u|, |v|, |w| \leq K$, there is a production $A_u \longrightarrow A_v A_w$. (Note that u, v, w may be 1 here). If A_v is a variable and $v = y$ is a relation holding in G, where $y \in X^{\pm 1}$, there is also a production $A_v \to y$. We take A_1 as the start symbol.

Given a word α in a derivation from A_1, replace every variable A_u occurring in α by u, to obtain a terminal string α'. By induction on the length of the derivation, $\alpha' = 1$ is a relation holding in G. Thus if $A_1 \xrightarrow{\bullet} w$, where w is a terminal string, then $w = 1$ is a relation holding in G, since $w' = w$. We have to prove the converse. First, we consider a diagonal triangulation of a polygon P, with a label in $(X^{\pm 1})^*$ assigned to each edge, such that:

(1) reading around the boundary of each triangle gives a word u with $\overline{\varphi}(u) = 1$
(2) if u is the label on any edge (including the boundary edges in P) then $|u| \leq K$.

(This need not be a K-triangulation as labels on boundary edges can have length greater than 1.) Let $y_1, \ldots, y_n$ be the labels on the edges of P reading anticlockwise round the boundary, starting at some vertex p, and let $w = y_1 \ldots y_n$. We show, by induction on the number of triangles, that $A_1 \xrightarrow{\bullet} \hat{w}$, where $\hat{w} = A_{y_1} \ldots A_{y_n}$.

If P is a 1-gon with label y_1, then $1 = y_1$ is a relation, so $A_1 \longrightarrow A_{y_1}$ is a production. If P is a bigon, $A_1 \longrightarrow A_{y_1} A_{y_2}$ is also a production. If P is a triangle, there is a derivation $A_1, A_{y_1} A_{y_1}^{-1}, A_{y_1} A_{y_2} A_{y_3}$.

Thus we can assume the number of triangles is at least two. By Lemma 5.18, there are two critical triangles. We can choose one whose two boundary edges do not meet at p, so $w = w_1 u v w_2$ where u, v are the labels on these boundary edges. The boundary of the triangle is labelled uvz^{-1} for some z, so $z = uv$ is a relation holding in G. This is illustrated in the following picture.

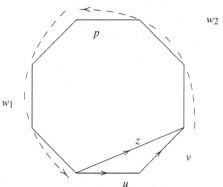

Figure 5.5

Removing the two edges on the boundary of the triangle gives a polygon with one less triangle and boundary label $w_1 z w_2$. By induction, $A_1 \xrightarrow{\bullet} \hat{w}_1 A_z \hat{w}_2$, and $A_z \longrightarrow A_u A_v$ is a production, so $A_1 \xrightarrow{\bullet} \hat{w}_1 A_u A_v \hat{w}_2 = \hat{w}$, completing the induction.

Now suppose $w = y_1 \ldots y_n$, with $y_i \in X^{\pm 1}$, $n \geq 1$ and $\overline{\varphi}(w) = 1$. There is a closed path α (starting at any chosen vertex) in $\Gamma(G, \varphi)$ with label w. There is a

K-triangulation of α with polygon P, say. Applying what has been proved to P, it follows that $A_1 \xrightarrow{\bullet} A_{y_1} \ldots A_{y_n}$. Using the productions $A_y \longrightarrow y$, we conclude that $A_1 \xrightarrow{\bullet} y_1 \ldots y_n = w$. Thus $L_E = W_\varphi(G) \setminus \{1\}$, and E is context-free, indeed in Chomsky normal form. By Cor. 1.2, $W_\varphi(G)$ is context-free.   □

Remark 5.1. Given a polygon P satisfying (1) and (2) in the second part of the proof, let e be an edge, given an orientation so it starts at a vertex a and ends at a vertex b. Let α be a sequence of boundary edges joining b to a, such that $e\alpha$ is a simple closed curve. Then reading around $e\alpha$ gives a word w with $\overline{\varphi}(w) = 1$. This can be proved by induction on the number of triangles of P, removing a critical triangle when the number of triangles is at least two. Details are left to the reader.

The next step is to show that an infinite context-free group has more than one end. For this purpose we do not need to define an end of a group, but only to give a suitable definition of the number of ends. A graph is called *locally finite* if there are only finitely many edges incident with every vertex. If Γ is a connected locally finite graph and F is a finite subgraph, let $\Gamma \setminus F$ be the graph obtained by removing all edges of F from Γ. Then $\Gamma \setminus F$ has only finitely many components, so only finitely many infinite components. (The components of a graph are the maximal connected subgraphs, and the graph is the disjoint union of its components; two vertices are in the same component if and only if there is a path from one to the other.) We define the *number of ends* of Γ to be

$$e(\Gamma) = \sup_F \ (\text{the number of infinite components of } F).$$

Thus $e(\Gamma)$ is either an integer or ∞. If $F_1 \subseteq F_2 \subseteq F_3 \ldots$ is a sequence of finite subgraphs of Γ such that $\bigcup_{i=1}^{\infty} F_i = \Gamma$, and c_n is the number of infinite components of $\Gamma \setminus F_n$, it is easy to see that $e(\Gamma) = \lim_{n \to \infty} c_n$.

If $\varphi : X \longrightarrow G$ is a map, where X is finite and $\varphi(X)$ generates G, we define the number of ends $e(G)$ of G to be $e(\Gamma(G, \varphi))$. It is true, though not obvious, that this is independent of the choice of φ. It is also true but not obvious that $e(G)$ is either 0, 1, 2 or ∞. (See [3, Section 2].) However, it is clear that $e(G) = 0$ if and only if G is finite.

Theorem 5.22. *If G is an infinite group with context-free word problem, then $e(G) > 1$.*

Proof. Let $\varphi : X \longrightarrow G$ be a mapping such that $\varphi(X)$ generates G and with X finite, and let $\Gamma = \Gamma(G, \varphi)$. For $n \geq 1$, let V_n be the set of all vertices v of Γ such that $d(1, v) < n$, where d is the path metric. Let F_n be the subgraph of Γ with vertex set V_n whose edges are all edges of Γ whose endpoints are in V_n. Since Γ is locally finite, F_n is finite.

Given a natural number n, as in Anisimov's Theorem there exists $g \in G$ such that if $w = y_1 \ldots y_m$ is a shortest word such that $\overline{\varphi}(w) = g$, then $m = |w| > n$. Then $y_1 \ldots y_n$ is a shortest word representing $\overline{\varphi}(y_1 \ldots y_n)$, otherwise we could find a shorter word than w representing g. Thus the path starting at 1 with label $y_1 \ldots y_n$ is a shortest path in Γ from 1 to $\overline{\varphi}(y_1 \ldots y_n)$. Take $n = 2i$, where $i \geq 1$ and let $p_i = \overline{\varphi}(y_1 \ldots y_i)$, $q_i = \overline{\varphi}(y_{i+1} \ldots y_{2i})$. Then $d(1, p_i) = i = d(p_i, p_i q_i)$ and $d(1, p_i q_i) = 2i$. Translating by p_i^{-1}, $d(p_i^{-1}, 1) = i = d(1, q_i)$ and $d(p_i^{-1}, q_i) = 2i$. We put $u_i = p_i^{-1}$, $v_i = q_i$.

By Theorem 5.21, there is a constant K such that every closed path in Γ can be K-triangulated. Choose $N > 3K/2$. We claim that, if u_i, v_i are as above, then u_i and v_i are in different components of $\Gamma \setminus F_N$, for $i \geq 1$. This implies $\Gamma \setminus F_N$ has at least two infinite components for all such N, because it has only finitely many components. It then follows that $e(\Gamma) > 1$, as required.

Suppose u_i and v_i are in the same component of $\Gamma \setminus F_N$. Let α be a path in Γ of minimal length from 1 to u_i, let γ be a path in Γ of minimal length from v_i to 1, and let β be a path in $\Gamma \setminus F_N$ from u_i to v_i. Then $\delta = \alpha \beta \gamma$ is a closed path in Γ. There is a K-triangulation of δ with polygon P. Reading around the boundary of P from a suitable point, the sequence of vertices encountered corresponds to the sequence of vertices of Γ passed through by δ. Thus every boundary vertex of P corresponds to a vertex of Γ. Also, ∂P is divided into three consecutive arcs, corresponding to α, β and γ. By Lemma 5.19, there is a triangle T having vertices on all three arcs, which define corresponding vertices of Γ, say a on α, b on β and c on γ.

Further, the label on an edge of T defines a path in Γ joining the two corresponding vertices of Γ, with length at most K. This follows easily from Remark 5.1. Thus the distance between any two of a, b, c is at most K. It follows that $d(1, a) \geq N - K$, otherwise $d(1, b) \leq d(1, a) + d(a, b) < (N - K) + K = N$, contradicting $b \in \Gamma \setminus F_N$. Also, $i = d(1, u_i) = d(1, a) + d(a, u_i)$, hence $d(a, u_i) \leq i - N + K$. Similarly, $d(c, v_i) \leq i - N + K$. But then

$$d(u_i, v_i) \leq d(u_i, a) + d(a, c) + d(c, v_i) \leq 2(i - N + K) + K = 2i + (3K - 2N) < 2i$$

as $N > 3K/2$. This contradicts $d(u_i, v_i) = 2i$.   □

The extra hypothesis needed by Muller and Schupp to prove a context-free group has a free subgroup of finite index is accessibility.

Definition. Let G be a finitely generated group. An *accessible series* for G is a series of subgroups

$$G = G_0 \geq G_1 \geq \ldots G_n$$

where each G_i is of the form $G_{i+1} *_K H$ or an HNN-extension $\langle t, G_{i+1} \mid tHt^{-1} = K \rangle$, where in each case K is finite. The *length* of the series is n.

Definition. A finitely generated group is *accessible* if there is an upper bound for the lengths of accessible series of G, and the least upper bound for these lengths is called the *accessibility length* of G.

Theorem 5.23. *If G has context-free word problem and is accessible, then G has a free subgroup of finite index.*

Proof. The proof is by induction on the accessibility length s of G. If $s = 0$, then G has no decomposition as a non-trivial free product with amalgamation or an HNN-extension with finite amalgamated or associated subgroups. But Stallings Structure

Theorem ([3, Theorem 3.1]) says that any group with more than one end does have such a decomposition. By Theorem 5.22, G must be finite. If $s > 0$, we can write $G = G_{i+1} *_K H$ or an HNN-extension $\langle t, G_{i+1} \mid tHt^{-1} = K \rangle$ with K finite, and G_{i+1}, and in the first case H, have accessibility length at most $s - 1$. By Lemma 5.12, G_{i+1} and, in the first case, H are finitely generated, and by Lemma 5.14, they are context-free. By induction they have free subgroups of finite index. It follows from [10] in the free product case, and from [24] in the HNN case, that G also has a free subgroup of finite index. □

Before stating the final characterisation of context-free groups, the following observation is needed.

Lemma 5.24. *If a finitely generated group G has context-free word problem, it is finitely presented.*

Proof. Let X be a finite set and $\varphi : X \longrightarrow G$ be a mapping such that $\varphi(X)$ generates G. Then $W_\varphi(G)$ is context-free, and there is an associated positive integer p given by the Pumping Lemma (Lemma 1.9). Suppose $z \in W_\varphi(G)$ and $|z| \geq p$. Then by Lemma 1.9, we can write $z = uvwxy$, where $|vwx| \leq p$, $vx \neq \varepsilon$ and for all $i \geq 0$, $uv^iwx^iy \in W_\varphi(G)$. In particular, $z' = uwy \in W_\varphi(G)$, and $|z'| < |z|$. Let $w' = w^{-1}vwx$, so

$$|w'| = |w^{-1}| + |vwx| = |w| + |vwx| \leq 2|vwx| \leq 2p.$$

Also, $z'y^{-1}w'y$ represents the same element of $F(X)$ as z, and $\overline{\varphi}(z'y^{-1}w'y) = \overline{\varphi}(z) = 1$, hence $w' \in W_\varphi(G)$. If $|z'| \geq p$, we can repeat this procedure with z' in place of z. This leads to a product $m = \prod_{i=1}^r y_i^{-1} r_i y_i$, where $r_i \in W_\varphi(G)$ and $|r_i| \leq 2p$, such that m and z represent the same element of $F(X)$. By Lemma 5.6, $z = 1$ is a consequence of the finite set $R = \{ r \in W_\varphi(G) \mid |r| \leq 2p \}$, so $\langle X | R \rangle$ is a finite presentation of G via φ. Note that the words w' and z' can be obtained from the Cayley graph by the "unstitching" process described above, using the diagram where the arrows

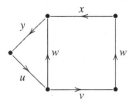

Figure 5.6

represent paths with the indicated labels.   □

Theorem 5.25. *Let G be a finitely generated group. The following are equivalent.*

(1) G has context-free word problem.
(2) G has a free subgroup of finite index.
(3) G has deterministic word problem.

Proof. Dunwoody [6] has shown that a finitely presented group is accessible, and it follows by Theorem 5.23 and Lemma 5.24 that (1) implies (2). We have seen that a finitely generated free group has deterministic word problem. It follows by Lemma 5.15 and Theorem A.5 that (2) implies (3). Obviously (3) implies (1).   □

There are other language classes we have not discussed, for which groups with word problem in the class have been studied. A *one-counter automaton* is a PDA

$$M = (Q, F, A, \{z, z_0\}, \tau, q_0, z_0)$$

where if $(q, a, z_0, q', \alpha) \in \tau$ then $\alpha \in \{z\}^* z_0$, and if $(q, a, z, q', \alpha) \in \tau$ then $\alpha \in \{z\}^*$. Starting in a configuration (q_0, w, z_0), the contents of the stack at any point (reading downwards) is $z^n z_0$, for some $n \in \mathbb{N}$. This is determined by n, so the stack is, in effect, a counter, which explains the name. The automaton is *deterministic* if it is deterministic as a PDA. A language L is *one-counter* if $L = L(M)$ for some one counter automaton M, and deterministic one-counter language is similarly defined. It has been shown in [13] that the following are equivalent, for a finitely generated group G.

(1) G has one-counter word problem.
(2) G has deterministic one-counter word problem.
(3) G has a cyclic subgroup of finite index (i.e. G is either finite or has an infinite cyclic subgroup of finite index).

(Note that Lemma 5.13 applies to the class of one-counter languages.)

A *real-time* language is a language in the class DTIME(n) (see the end of Chap. 3). As noted at the end of Chap. 1, the context-sensitive languages are those recognised by a linear bounded automaton, and these coincide with the languages in NSPACE(n) (see Theorem 12.2 and the remark following it in [21]). By Theorem 12.10 in [21], DTIME(n) $\subseteq$ DSPACE(n) $\subseteq$ NSPACE(n), so a real-time language is context-sensitive. A deterministic FSA may be viewed as a deterministic TM of time complexity n with one tape, so a regular language is a real-time language.

There has been some progress in studying groups whose word problem is a real-time language. It is proved in [16] that finitely generated nilpotent groups, word hyperbolic groups and geometrically finite groups have real-time word problem. In [18], the number of tapes needed by a TM of time complexity n to recognise the word problem of certain groups is investigated.

Another class, introduced by Aho, is the class of indexed languages, which are defined by "indexed grammars" and are recognised by "one-way nested stack automata". Further details are contained in §14.3 and the bibliographic notes at the end of Chap. 14 in [21]. (Indexed grammars are also defined in [2].) This class lies between the context-free and context-sensitive language classes. Nevertheless, it is an open problem whether or not the class of groups with indexed word problem coincides with the class of groups having context-free word problem, that is, the finitely generated groups with a free subgroup of finite index.

A *simple* grammar is one in Greibach normal form (see Theorem 4.10), such that for every variable A and terminal a, there is at most one string α such that $A \longrightarrow a\alpha$ is a production. A language is *simple* if it can be generated by a simple grammar.

Given a simple grammar, the corresponding PDA in the proof of Theorem 4.14 is deterministic, so a simple language is strict deterministic, hence prefix-free (Remark 4.2). We also allow a grammar with the single production $S \longrightarrow \varepsilon$ as a simple grammar, generating the language $\{\varepsilon\}$, which is also strict deterministic (Remark 4.2).

Let $\varphi : X \longrightarrow G$ be a mapping such that $\varphi(X)$ generates the group G, and X is finite. Now $W_\varphi(G)$ is prefix-free only in the extreme case $X = \emptyset$, when G is trivial. For otherwise, $W_\varphi(G)$ will contain a word xx^{-1}, where $x \in X$, and its prefix ε. Instead, consider the *reduced word problem*, which is the set $R_\varphi(G)$ whose members are those non-empty words $w \in W_\varphi(G)$ such that no prefix of w, other than ε and w, is in $W_\varphi(G)$. Also, let $I_\varphi(G)$ (the *irreducible word problem*) be the set of non-empty words in $W_\varphi(G)$ which have no subword, other than ε and w, in $W_\varphi(G)$, a subset of $R_\varphi(G)$.

Haring-Smith [11] has shown that $R_\varphi(G)$ is simple for some φ if and only if G is *plain*, that is, a free product of a finitely generated free group and finitely many finite groups. There is again a characterisation involving the Cayley graph; $R_\varphi(G)$ is simple if and only if $\Gamma(G, \varphi)$ has the property that there are only finitely many circuits passing through any vertex. This is equivalent to saying that $I_\varphi(G)$ is finite. Haring-Smith conjectured that $R_\varphi(G)$ is strict deterministic for some φ if and only if G has a plain subgroup of finite index. This was proved in [30], in fact $R_\varphi(G)$ is strict deterministic for some φ if and only if G has context-free word problem. (Note that, by [10], a plain group has a free subgroup of finite index, so we see directly that having a plain subgroup of finite index is equivalent to having a finitely generated free subgroup of finite index.)

Another variant of $W_\varphi(G)$ is its complement, $(X^{\pm 1})^* \setminus W_\varphi(G)$, which is called the co-word problem. Groups with context-free co-word problem are considered in [17], and groups with indexed co-word problem have been studied by D. Holt and C. Röver [19].

Automatic Groups

Let X be a set of monoid generators for a group G. That is, there is a mapping $\varphi : X \to G$ such that the extension $\overline{\varphi} : (X^{\pm 1})^* \to G$ maps X^* onto G. Assume X is finite.

Definition. Let L be a language with alphabet X. Then (X, L) is called a *rational structure* for G if L is regular and $\overline{\varphi}(L) = G$.

Choose a letter $\$ \notin X$ (the "padding symbol"). Define $X' = X \cup \{\$\}$ and

$$X(2, \$) = (X' \times X') \setminus \{(\$, \$)\}.$$

Now define $\mu : X^* \times X^* \to X(2, \$)^*$ by: if $u = x_1 \ldots x_m$, $v = y_1 \ldots y_n \in X^*$, then

$$\mu(u, v) = \begin{cases} (x_1, y_1) \ldots (x_n, y_n)(x_{n+1}, \$) \ldots (x_m, \$) & \text{if } m > n \\ (x_1, y_1) \ldots (x_m, y_m) & \text{if } m = n \\ (x_1, y_1) \ldots (x_m, y_m)(\$, y_{n+1}) \ldots (\$, y_n) & \text{if } m < n \end{cases}$$

Let (X,L) be a rational structure for G. For $w \in X^*$, define

$$L_w = \{\mu(w_1, w_2) \mid w_1, w_2 \in L \text{ and } \overline{\varphi}(w_1) = \overline{\varphi}(w_2 w)\}.$$

Definition. The rational structure (X,L) is an *automatic structure* for G if L_ε and L_x (for all $x \in X$) are regular languages. A group is *automatic* if it has an automatic structure.

Examples of automatic groups are word hyperbolic groups (in particular, finite groups and finitely generated free groups), finitely generated abelian groups, braid groups and many 3-manifold groups. (See [7], Theorem 3.4.5, Theorem 4.3.1, Chapter 9 and Chapter 12.) Also, many small cancellation groups are automatic (see [8].)

Our intention is to give a characterisation of automatic groups in terms of the Cayley graph. First, given the automatic structure (X,L) on G (via $\varphi : X \to G$), we shall construct finite state automata M_x, for $x \in X \cup \{\varepsilon\}$, which under certain circumstances will recognise L_x. To do this, we let B be a finite subset of G containing 1.

The language $L\{\$\}^*$ is regular by Lemma 1.5, so is recognised by a deterministic FSA, say M (with tape alphabet $X \cup \{\$\}$). Let δ be the transition function of M. The FSA M_x has set of states $Q \times Q \times B$, where Q is the set of states of M. The tape alphabet is $X(2,\$)$, and the initial state is $(q_0, q_0, 1)$, where q_0 is the initial state of M. We modify M_x by identifying all states of the form (q_1, q_2, g), where either q_1 or q_2 is a dead state of M, to a single state f. (A state q of a FSA is called *dead* if it is not a final state and there is no path in the transition diagram from q to a final state. For an example, see Example (2) of a transition diagram in Chapter 1.) Now given a state $p = (q_1, q_2, g)$ and $a = (y_1, y_2) \in X(2,\$)$, there is a transition (p, a, p'), where

$$p' = \begin{cases} (\delta(q_1, y_1), \delta(q_2, y_2), h) & \text{if } h \in B \\ f & \text{otherwise} \end{cases}$$

where $h = \overline{\varphi}(y_2)^{-1} g \overline{\varphi}(y_1)$ and $\overline{\varphi}(\$)$ is defined to be 1_G. $y_1 = \$$, y_1^{-1} is replaced by 1, and if $y_2 = \$$, y_2 is replaced by 1. The final states of M_x are those of the form $(q_1, q_2, \overline{\varphi}(x))$, where q_1, q_2 are final states of M. The FSA M_x is called a *standard automaton* based on M and B. The reason for this strange definition of M_x will become apparent when it is used.

Before giving our characterisation of automatic structures, a definition is needed. Again we assume (X,L) is an automatic structure on G (via $\varphi : X \to G$), and let d be the path metric in the Cayley graph $\Gamma(G, \varphi)$. Let $w = a_1 \ldots a_n \in X^*$; for $t \in \mathbb{N}$, put

$$w(t) = \begin{cases} a_1 \ldots a_t & \text{if } t \leq n \\ a_1 \ldots a_n & \text{if } t > n \end{cases}$$

Definition. Let K be a positive real number. Two words u, $v \in X^*$ are called *K-fellow travellers* if, for all $t \in \mathbb{N}$,

$$d(\overline{\varphi}(u(t)), \overline{\varphi}(v(t))) \leq K.$$

Informally, the paths with labels $u(t)$, $v(t)$ starting at 1_G are uniformly close in the metric d.

Theorem 5.26. *A rational structure for a group G (via $\varphi : X \to G$) is an automatic structure for G if and only if there exists $K > 0$ such that, for all u, $v \in L$ and $x \in X \cup \{\varepsilon\}$, if $\overline{\varphi}(u) = \overline{\varphi}(vx)$ then u and v are K-fellow travellers.*

Proof. Assume (X,L) is an automatic structure for G. Let M_x be a FSA recognising L_x, for $x \in X \cup \{\varepsilon\}$. Let N be the maximum number of states in any of the automata M_x. If u, $v \in L$ and $\overline{\varphi}(u) = \overline{\varphi}(vx)$ then M_x accepts $\mu(u,v)$. Let $t \in \mathbb{N}$. After reading the prefix $\mu(u(t),v(t))$ of $\mu(u,v)$, suppose M_x is in state q. Then there is a path (in the transition diagram of M_x) from q to a final state of M_x, with label $\mu(w,z)$, where $u = u(t)w$, $v = v(t)z$. Take a shortest path from q to this final state, with label $\mu(w',z')$ say. This path never visits the same vertex twice (otherwise we could shorten it) so has length $|\mu(w',z')| \le N - 1$. Then M_x accepts $\mu(u(t)w',v(t)z')$, so $\overline{\varphi}(u(t)w') = \overline{\varphi}(v(t)z'x)$. Hence, there are paths in the Cayley diagram as illustrated:

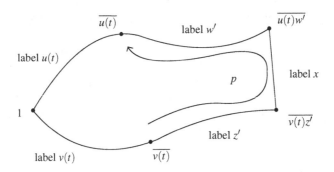

Figure 5.7

(where $\overline{u(t)} = \overline{\varphi}(u(t))$, etc). Thus $d(\overline{u(t)},\overline{v(t)}) \le$ length of the path $p = |w'| + |z'| + 1$ and $|w'|$, $|z'| \le |\mu(w',z')| \le N - 1$, so $d(\overline{u(t)},\overline{v(t)}) \le 2N - 1$. We can take $K = 2N - 1$.

Conversely, assume K exists as in the theorem. Let $B = \{g \in G \mid d(1,g) \le K\}$ and let M_x ($x \in X \cup \{\varepsilon\}$) be the standard automaton corresponding to B and a deterministic FSA M recognising $L\{\$\}^*$ constructed above. We claim that M_x recognises L_x, hence (X,L) is an automatic structure for G.

Let $x \in X \cup \{\varepsilon\}$ and suppose $(w_1,w_2) \in X^* \times X^*$, where $w_1, w_2 \in L$ and $\overline{w_1} = \overline{w_2 x}$. Then

$$d(1,\overline{w_2(t)}^{-1}\overline{w_1(t)}) = d(\overline{w_2(t)},\overline{w_1(t)}) \le K$$

for all $t \in \mathbb{N}$. If $w_1(t+1) = w_1(t)y_1$ and $w_2(t+1) = w_2(t)y_2$, and $g = \overline{w_2(t)}^{-1}\overline{w_1(t)}$, $h = \overline{w_2(t+1)}^{-1}\overline{w_1(t+1)}$, then $h = \overline{\varphi}(y_2)^{-1}g\overline{\varphi}(y_1)$, and g, $h \in B$. This is illustrated by a picture representing part of the Cayley graph.

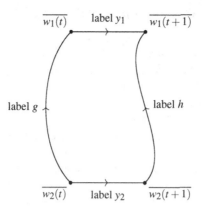

Figure 5.8

(Note that $\overline{w_1(t)}$ may be equal to $\overline{w_1(t+1)}$, when the edge with label y_1 is absent and $\overline{\varphi}(y_1)$ means 1. Similarly the bottom edge may be missing.) It follows from the construction of M_x that M_x accepts $\mu(w_1, w_2)$. Conversely, suppose there is a computation of M_x with label $\mu(w_1, w_2)$ ending at a state (q_1, q_2, g). An easy induction on the length of the computation shows that $g = \overline{\varphi}(w_2)^{-1}\overline{\varphi}(w_1)$, and there are computations of M with labels $w_1\$^k$, $w_2\$^l$ for some $k, l \geq 0$, ending respectively at q_1, q_2. Therefore if (q_1, q_2, g) is a final state, $w_1\$^k$, $w_2\$^l \in L\{\$\}^*$, so w_1, $w_2 \in L$, and $g = \overline{\varphi}(x)$, hence $\overline{\varphi}(w_1) = \overline{\varphi}(w_2 x)$, so $\mu(w_1, w_2) \in L_x$. This completes the proof. □

More can be gleaned from the first paragraph of the proof. Suppose $|u| > |v| + N$. Take $t = |v|$. The path from q to a final state visits some vertex twice, so can be shortened. Then $\overline{\varphi}(u(t)w') = \overline{\varphi}(vx) = \overline{\varphi}(u)$, $u(t)w' \in L$ and $|u(t)w'| < |u|$. Similarly if $|v| > |u| + N$, there is a shorter element of L representing the same element of G as v. This leads to the following lemma.

Lemma 5.27. *Let (X, L) be an automatic structure for a group G via φ. There is a positive integer N such that if $w \in L$ and g is a vertex of $\Gamma(G, \varphi)$ at distance at most 1 from $\overline{\varphi}(w)$, then $g = \overline{\varphi}(u)$ for some $u \in L$ of length at most $|w| + N$.*

Proof. Let w' be a representative of g in L. Either $w = w'x$ or $w' = wx$ for some $x \in X \cup \{\varepsilon\}$. Take N as in the proof of Theorem 5.26. By the observations preceding the lemma, if $|w'| > |w| + N$, it can be replaced by a shorter word, and the lemma follows. □

Suppose $G = \langle X \mid R \rangle^\varphi$. Recall that, if $\overline{\varphi}(w) = 1$, then $w =_{F(X)} \prod_{i=1}^{k} u_i r_i^{\pm 1} u_i^{-1}$ for some $u_i \in F(X)$, $r_i \in R$, $k \in \mathbb{N}$ (see Cor. 5.7). We put

$$a(w) = \text{the least possible value of } k.$$

Definition. The *isoperimetric function f* of the presentation is the mapping $f : \mathbb{N} \to \mathbb{N}$ given by

$$f(n) = \max\{a(w) \mid |w| \leq n, \overline{\varphi}(w) = 1\}.$$

Lemma 5.28. *Let $\langle X \mid R \rangle$ be a finite presentation of a group G, via φ. The following are equivalent.*

(1) *The isoperimetric function is bounded above by a recursive function $\mathbb{N} \to \mathbb{N}$.*
(2) *G has solvable word problem.*
(3) *The isoperimetric function is recursive.*

Proof. We shall not prove this, but refer to [7, Theorem 2.2.5]. This depends on the previous result [7, Theorem 2.2.4], which gives a bound on the lengths of the u_i, when a word $w \in W_\varphi(G)$ is written as in Equation $(**)$ in Cor. 5.7. The proof of this is geometric, and it would be too great a digression to prove it. It uses van Kampen diagrams, which are discussed in Chapter V of [25]. (They are based on the idea, used several times in this chapter, of representing part of the Cayley graph by a diagram in the plane.) Also, to convert the argument of [7, Theorem 2.2.5] into a precise form, showing that (1) and (3) are equivalent to the statement that $W_\varphi(G)$ is recursive, is a tedious, and possibly futile, exercise. $\square$

Theorem 5.29. *Suppose G is automatic. Then*

(1) *G has a finite presentation whose isoperimetric function is bounded above by a quadratic function*
(2) *G has solvable word problem.*

Proof. Let (X,L) be an automatic structure for G via φ, and let K be as in Theorem 5.26. Let $w \in (X^{\pm 1})^*$, say $w = y_1 \ldots y_n$, put $g_i = \overline{\varphi}(y_1 \ldots y_i)$ $(0 \leq i \leq n)$ and let w_i be an element of L, of shortest possible length, such that $\overline{\varphi}(w_i) = g_i$. There is a path p_i in $\Gamma(G,\varphi)$ from 1 to g_i with label w_i, and an edge e_i from g_i to g_{i+1} with label y_{i+1}, for $0 \leq i \leq n-1$. The situation is illustrated in the planar diagram on the next page. The top curve represents p_i and the bottom curve p_{i+1}. The straight lines represent paths of length at most K. These exist because by Theorem 5.26, w_i and w_{i+1} are K-fellow travellers. Again denoting $\overline{\varphi}(w_i(t))$ by $\overline{w_i(t)}$, etc., the closed path starting at $\overline{w_i(t)}$ and passing through $\overline{w_{i+1}(t)}$, $\overline{w_{i+1}(t_{i+1})}$ and $\overline{w_i(t+1)}$ has length at most $2K+2$. (Note that, for sufficiently large t, $\overline{w_i(t)}$ and $\overline{w_i(t+1)}$ may coincide; this happens when $|w_{i+1}| > |w_i|$. Similarly $\overline{w_{i+1}(t)}$ and $\overline{w_{i+1}(t+1)}$ may coincide.) Thus the closed path $p_i e_i \bar{p}_{i+1}$ has been decomposed into $\max\{|w_i|,|w_{i+1}|\}$ closed paths of length at most $2K+2$. Let h_i be the label on $p_i e_i \bar{p}_{i+1}$.

Unstitching the picture as previously described, h_i is equal in $F(X)$ to a product of $\max\{|w_i|,|w_{i+1}|\}$ conjugates of the form uru^{-1}, where $\overline{\varphi}(r) = 1$ and $|r| \leq 2K+2$. Now if $\overline{\varphi}(w) = 1$, $\overline{\varphi}(w_n) = g_n = 1$, so we can take $w_n = w_0$, and then w itself is conjugate in $F(X)$ to $h_0 \ldots h_{n-1}$, so is equal in $F(X)$ to a product of conjugates of elements of the finite set $R = \{r \in (X^{\pm 1})^* \mid \overline{\varphi}(r) = 1 \text{ and } |r| \leq 2K+2\}$. Thus w is a consequence of R, so $G = \langle X \mid R \rangle^\varphi$.

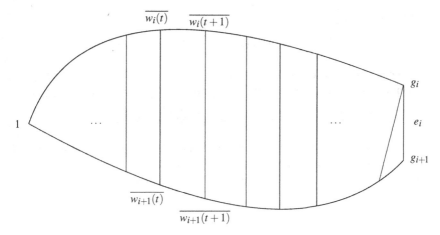

Figure 5.9

Let $n_0 = |w_0|$. Inductively, using Lemma 5.27, we can find a positive integer N and $u_i \in L$ such that $\overline{\varphi}(u_i) = \overline{\varphi}(w_i)$ and $|u_i| \le n_0 + iN$ for $0 \le i \le n$. Since the w_i were chosen of minimal length, $|w_i| \le n_0 + nN$ for all i. Hence, h_i is a product of at most $n_0 + nN$ conjugates of elements of R, and so w is a product of at most $n(n_0 + nN)$ conjugates of elements of R. This proves (1). Now (2) follows from Lemma 5.28 and (1). $\qquad\square$

In fact, it is known that a finitely generated subgroup of an automatic group has deterministic context-sensitive word problem ([36]). For further reading on the basics of automatic groups, see [7] and [37]. There is a useful generalisation to the notion of a group having an asynchronous $\mathscr{A}$-combing ([2]). Here $\mathscr{A}$ is a "full abstract family of languages", which is a class of languages closed under certain operations, most of which we have encountered. A group has an asynchronous regular combing if and only it is asynchronously automatic ([7, Chap. 7]). The idea of asynchronously automatic group generalises that of automatic group.

Exercises on Chapter 5

In the first three questions, suppress the mapping φ in the definition of "presentation via φ", as in the example on p. 96.

1. Show that the following are presentations of the trivial group.

(a) $\langle x, y \mid x^2 = y^3, xyx = yxy \rangle$.
(b) $\langle x, y \mid xyx^{-1} = y^2, yxy^{-1} = x^2 \rangle$.
(c) $\langle x, y, z \mid xyx^{-1} = y^2, yzy^{-1} = z^2, zxz^{-1} = x^2 \rangle$.

2. Show that

$$\langle x_1,\ldots,x_{n-1} \mid x_i^2 = 1 \ (i \leq n-1), (x_i x_{i+1})^3 = 1 \ (i \leq n-2), (x_i x_j)^2 = 1 \ (j < i-1)\rangle$$

is a presentation of the symmetric group S_n. (Hint: consider H, the subgroup generated by $x_1,\ldots,x_{n-2}$, and the set of cosets

$$\{H, Hx_{n-1}, Hx_{n-1}x_{n-2}, \ldots, Hx_{n-1}x_{n-2}\ldots x_2 x_1\}$$

and use the method in the example of a presentation of S_3 given in the text. An induction on n is needed.)

3. Show that $\langle x_1,\ldots,x_{n-2} \mid R\rangle$, where R is the set of relations

$$\{x_1^3 = x_i^2 = 1 \ (2 \leq i \leq n-2), (x_i x_{i+1})^3 = 1 \ (i \leq n-3), (x_i x_j)^2 = 1 \ (j < i-1)\}$$

is a presentation of the alternating group A_n for $n \geq 3$. (Hint: this is similar to the previous exercise: consider H, the subgroup generated by $x_1,\ldots,x_{n-3}$, and the set of cosets

$$\{H, Hx_{n-2}, Hx_{n-2}x_{n-3}, \ldots, Hx_{n-2}x_{n-3}\ldots x_2 x_1\}, Hx_{n-2}x_{n-3}\ldots x_2 x_1^2.)$$

4. Let X be a subset of a group G. Prove that the following are equivalent.

(a) The extension of the inclusion mapping $X \longrightarrow G$ to a group homomorphism $F(X) \longrightarrow G$ given by Lemma 5.5 is an isomorphism.

(b) Given any mapping $\alpha : X \longrightarrow H$, where H is a group, there is a unique extension of α to a homomorphism $G \longrightarrow H$.

(c) X generates G, and no non-empty reduced word in $(X^{\pm 1})^*$ represents the identity element of G.

(When these conditions are satisfied, G is said to be free with basis X.)

5. Suppose F_1 is free with basis X_1 and F_2 is free with basis X_2. Show that F_1 is isomorphic to F_2 if and only if $|X_1| = |X_2|$, where $|X_i|$ is the cardinality of X_i. (Hint: if F_i^2 is the subgroup generated by $\{u^2 \mid u \in F_i\}$, then F_i^2 is a normal subgroup of F_i and the quotient is a vector space over the field of two elements, with basis the image of X_i. If you are unfamiliar with infinite cardinals and infinite dimensional vector spaces, assume X_i is finite, so $|X_i|$ is the number of elements in X_i, for $i = 1, 2$.) Thus, if F is free with basis X, we define the *rank* of F to be $|X|$.

6. If F is a finitely generated free group, show that F has a finite basis.

7. Let F be free with basis $\{x,y\}$. Show that the set $Y = \{x^i y x^{-i} \mid i \in \mathbb{N}\}$ is a basis for the subgroup of F it generates. (Hint: take a reduced word in $(Y^{\pm 1})$, say $u_1\ldots u_n$, where $u_j = x^{i_j} y^{e_j} x^{-i_j}$ $(i_j \in \mathbb{N}, e_j = \pm 1)$, which represents an element g of F. Show that the reduced word in $\{x,y\}^{\pm 1}$ representing g has the form $x^{i_1} y^{e_1} v_1 y^{e_2} v_2 \ldots v_{n-1} y^{e_n} x^{-i_n}$, where each v_i is a power of x or x^{-1} (possibly empty), by induction on n. Now use Exercise 4.) Thus a free group of rank 2 contains a free group of countably infinite rank.

8. Show that a group is a free group if and only if it is isomorphic to a free product of infinite cyclic groups.

9. Using a suitable set of generators, describe the Cayley graph of the following groups, and hence determine the number of ends of the group.

(a) The free group of rank 2.

(b) The free abelian group of rank 2.

(c) The symmetric group of degree 3.

("Suitable" means a basis, in the appropriate sense, in (a) and (b), and in (c), a 3-cycle and a transposition. A good way to describe the graphs is to draw enough of them to indicate the general structure of the graph. Detailed proofs for the number of ends are not required.)

Appendix A
Results and Proofs Omitted in the Text

We begin with the assertion at the beginning of Chapter 1, that a type 1 language can be generated by a grammar whose productions are context-sensitive.

Note that, at this point in Chapter 1, $S \longrightarrow \varepsilon$ is not allowed as a production in a type 1 grammar. However Lemma A.2 below is true if modified to allow it, adding "except possibly $S \longrightarrow \varepsilon$". This is because the arguments in Lemma 1.1 and Cor. 1.2 apply.

Lemma A.1. *If $G = (V_N, V_T, P, S)$ is a grammar of type 0 or 1, then $L_G = L_{G'}$ for some grammar G' of the same type, such that all productions of G' are either of the form $\alpha \longrightarrow \beta$, where α, β are strings of non-terminal symbols, or of the form $A \rightarrow a$, where A is a non-terminal symbol and a is a terminal symbol.*

Proof. For every $a \in V_T$, take a new letter X_a. Let $G' = (V_N', V_T, P', S)$, where $V_N' = V_N \cup \{X_a \mid a \in V_T\}$ and P' consists of:

$$X_a \longrightarrow a \qquad \text{for } a \in V_T$$
and $\quad \alpha' \longrightarrow \beta' \qquad$ for $\alpha \longrightarrow \beta$ in P, where α' and β' are obtained from α, β by replacing every occurrence of a letter $a \in V_T$ by X_a.

Then G' is of the same type as G, and is the required grammar. For if $\gamma \in L_G$, modify a G-derivation of γ from S, by replacing every production $\alpha \longrightarrow \beta$ used by $\alpha' \longrightarrow \beta'$, to obtain a G'-derivation of γ'. Then by use of the productions $X_a \longrightarrow a$, we obtain a G'-derivation of γ. Hence $L_G \subseteq L_{G'}$.

Conversely, given a G'-derivation of $\gamma \in L_{G'}$, when a production $X_a \longrightarrow a$ is used, the resulting occurrence of a is never changed. Move this production to the end of the derivation, replacing the occurrence of a by X_a until the production $X_a \longrightarrow a$ is used. Repeating this procedure, we obtain a G'-derivation of γ in which all uses of productions $\alpha' \longrightarrow \beta'$ occur first. This produces a word in the new letters X_a which is then converted to γ, so this word must be γ'. Now replace every use of a production $\alpha' \longrightarrow \beta'$ by the production $\alpha \longrightarrow \beta$ and delete all uses of productions $X_a \longrightarrow a$ at the end. This gives a G-derivation of γ from S. Hence $L_G = L_{G'}$. $\qquad \square$

Lemma A.2. *Let G be a type 1 grammar. Then $L_G = L_{G'}$ for some grammar G' in which all productions are context-sensitive.*

Proof. We can assume the productions of G are as in Lemma A.1. Given a production $a_1 \ldots a_n \longrightarrow b_1 \ldots b_m$ ($m \geq n$, a_i, b_i non-terminal letters) which is not context-sensitive (so $n > 1$), modify G as follows. Add new non-terminal letters $A_1, \ldots A_n$ and $B_1, \ldots B_m$ (distinct, even if the a_i, b_i aren't), then replace this production by the productions:

$$a_1 \ldots a_n \longrightarrow a_1 \ldots a_{n-1} A_n \tag{1}$$

$$a_1 \ldots a_{n-1} A_n \longrightarrow a_1 \ldots a_{n-2} A_{n-1} A_n \tag{2}$$

$$\vdots \qquad\qquad\qquad\qquad\qquad \vdots$$

$$a_1 A_2 \ldots A_n \longrightarrow A_1 \ldots A_n \tag{n}$$

$$A_1 \ldots A_n \longrightarrow A_1 \ldots A_{n-1} B_n \ldots B_m \tag{n+1}$$

$$A_1 \ldots A_{n-1} B_n \ldots B_m \longrightarrow A_1 \ldots A_{n-2} B_{n-1} B_n \ldots B_m \tag{n+2}$$

$$\vdots \qquad\qquad\qquad\qquad\qquad \vdots$$

$$A_1 B_2 \ldots B_m \longrightarrow B_1 B_2 \ldots B_m \tag{2n}$$

$$B_1 B_2 \ldots B_m \longrightarrow b_1 B_2 \ldots B_m \tag{2n+1}$$

$$b_1 B_2 \ldots B_m \longrightarrow b_1 b_2 B_3 \ldots B_m \tag{2n+2}$$

$$\vdots \qquad\qquad\qquad\qquad\qquad \vdots$$

$$b_1 \ldots b_{m-1} B_m \longrightarrow b_1 \ldots b_m \tag{2n+m}$$

(The reader should check that these are context-sensitive.)

Call the new grammar G_1. Any use of the old production can be replaced by using these $2n + m$ productions in succession, hence $L_G \subseteq L_{G_1}$. Suppose $\alpha \in L_{G_1}$ and there is a G_1-derivation of α from S (the start symbol) using the new productions. The first time such a production is used, it must be (1), since up to that point none of the new non-terminal letters have appeared. This introduces A_n, and this occurrence of A_n must eventually be changed by use of a production (α is a string of terminal letters). This can only be done by use of (2).

The reason is that the letter to the left of A_n is either from the original alphabet, or the right-hand letter of the right-hand side of a new production, which can only be A_n or B_m. Similarly, the letter to the right of A_n (if any) is either from the original alphabet, or A_1 or B_1. But no word on the left-hand side of the new productions contains any of the words $A_n A_n$, $B_m A_n$, $A_n A_1$ or $A_n B_1$. Thus (2) is the only production that can be used, so part of the derivation has the form:

$$\ldots, u_1 a_1 \ldots a_n v_1, u_1 a_1 \ldots A_n v_1, \ldots, u_1' a_1 \ldots A_n v_1', u_1' a_1 \ldots A_{n-1} A_n v_1', \ldots$$

which can be replaced by

$$\ldots, u_1 a_1 \ldots a_n v_1, \ldots, u_1' a_1 \ldots a_n v_1', u_1' a_1 \ldots A_n v_1', u_1' a_1 \ldots A_{n-1} A_n v_1', \ldots .$$

(The use of (1) is moved until just before the use of (2).) Similarly, the next time $A_{n-1}A_n$ is changed by a production, it must be by use of (3), and we can change the derivation so (1), (2) and (3) are used in succession. Eventually we obtain a G_1-derivation in which (1) through (2n+m) are used in succession, to change an occurrence of the string $a_1 \ldots a_n$ to $b_1 \ldots b_m$. These can be replaced by a single use of the original production.

Continuing, we eventually remove all use of the new productions, giving a G-derivation of α (the fact that the original productions are now used does not affect the argument). Hence $L_{G_1} = L_G$. Finally, repetition of the procedure replacing G by G_1 will remove all productions which are not context-sensitive, giving the required grammar G'.   □

The next result to be proved is Lemma 2.18. It is necessary to read the relevant part of Chapter 2 to understand the statement and proof.

Lemma 2.18. There is a primitive recursive function $Next : \mathbb{N} \to \mathbb{N}$ such that

$$Next(Code(c)) = Code(\delta(c))$$

for all $c \in C'$.

Proof. Let $c = (q, a, \alpha, \beta)$, so $Code(c) = 2^q 3^a 5^{\sigma(\alpha)} 7^{\sigma(\beta)}$. Put $x = Code(c)$, and use Lemma 1.10. To simplify notation, we omit subscripts and write R, N, D instead of $R_{T'}, N_{T'}, D_{T'}$.

(1) If $D(q, a) = 0$, $Code(\delta(c)) = 2^{N(q,a)} 3^{\beta(0)} 5^{\sigma(\alpha')} 7^{\sigma(\beta')}$ and

$$
\begin{aligned}
N(q, a) &= N(\log_2(x), \log_3(x)) \\
\beta(0) &= \mathrm{rem}(2, \log_7(x)) \\
\sigma(\alpha') &= R(q, a) + 2\alpha(0) + 2^2\alpha(1) + \ldots = R(\log_2(x), \log_3(x)) + 2\log_5(x) \\
\sigma(\beta') &= \beta(1) + 2\beta(2) + \ldots = \mathrm{quo}(2, \sigma(\beta)) = \mathrm{quo}(2, \log_7(x))
\end{aligned}
$$

(2) If $D(q, a) = 1$, $Code(\delta(c)) = 2^{N(q,a)} 3^{\alpha(0)} 5^{\sigma(\alpha')} 7^{\sigma(\beta')}$ and similarly

$$
\begin{aligned}
N(q, a) &= N(\log_2(x), \log_3(x)) \\
\alpha(0) &= \mathrm{rem}(2, \log_5(x)) \\
\sigma(\alpha') &= \mathrm{quo}(2, \log_5(x)) \\
\sigma(\beta') &= R(\log_2(x), \log_3(x)) + 2\log_7(x)
\end{aligned}
$$

Hence, we define $Next(x) = 2^{F_1(x)} 3^{F_2(x)} 5^{F_3(x)} 7^{F_4(x)}$ (for any $x \in \mathbb{N}$) where, putting $E(x) = D(\log_2(x), \log_3(x))$:

$$
\begin{aligned}
F_1(x) &= N(\log_2(x), \log_3(x)) \\
F_2(x) &= (1 \dot{-} E(x))\mathrm{rem}(2, \log_7(x)) + E(x)\mathrm{rem}(2, \log_5(x)) \\
F_3(x) &= (1 \dot{-} E(x))(R(\log_2(x), \log_3(x)) + 2\log_5(x)) + E(x)\mathrm{quo}(2, \log_5(x)) \\
F_4(x) &= (1 \dot{-} E(x))\mathrm{quo}(2, \log_7(x)) + E(x)(R(\log_2(x), \log_3(x)) + 2\log_7(x))
\end{aligned}
$$

Since $F_1, \ldots, F_4$ are primitive recursive, so is *Next*.   □

Now we prove two lemmas on deterministic pushdown automata which are needed at the end of Chapter 4, where these and related ideas are defined. We shall need another definition concerning them.

Definition. A deterministic PDA $M = (Q, F, A, \Gamma, \tau, q_0, z_0)$ is said to *always scan its entire input* if, for all $w \in A^*$, $(q_0, w, z_0) \xrightarrow{M} (q, \varepsilon, \gamma)$ for some $q \in Q$ and $\gamma \in \Gamma^*$.

Also, we shall describe a transition of a PDA starting $(q, \varepsilon, \ldots)$ as an ε-transition.

The ways in which M can fail to always scan its entire input are firstly, that it halts without reading the entire word on the tape. This can happen if M empties its stack, or if there is no transition beginning (q, a, z) or (q, ε, z), where M is in state q, a is the next letter on the tape and z is on top of the stack. Secondly, it can happen that M continues indefinitely to use ε-transitions without reading another letter from the tape. This observation is the basis for the construction in the next lemma. This makes use of a state d (the "dead state") to continue to read from the tape when any of the situations above is encountered. There is also an extra final state f to accept any words that M accepts in the second situation, when it continues indefinitely to use ε-transitions. Such a word will be a proper prefix of the word on the tape. (The proper prefixes of a word w are the prefixes of w other than w itself.)

Lemma A.3. *If L is a deterministic language, then $L = L(M')$ for some deterministic PDA M' which always scans its entire input.*

Proof. There is a deterministic PDA $M = (Q, F, A, \Gamma, \tau, q_0, z_0)$ such that $L = L(M)$. There is a new PDA $M' = ((Q \cup \{q_0', d, f\}, F \cup \{f\}, A, \Gamma \cup \{x_0\}, \tau', q_0', x_0)$, where τ' is defined as follows.

(1) $(q_0', \varepsilon, x_0, q_0, z_0 x_0) \in \tau'$.
(2) For all $q \in Q$, $a \in A$ and $z \in \Gamma$, if no transition in τ starts with (q, a, z) or (q, ε, z), then $(q, a, z, d, z) \in \tau'$.
(3) For all $q \in Q$, $a \in A$, $(q, a, x_0, d, x_0) \in \tau'$.
(4) For all $a \in A$, $z \in \Gamma \cup \{x_0\}$, $(d, a, z, d, z) \in \tau'$.
(5) If there is an infinite sequence of configurations of M:

$$(q, \varepsilon, z), (q_1, \varepsilon, \gamma_1), (q_2, \varepsilon, \gamma_2), \ldots$$

where $z \in \Gamma$ and each configuration $(q_i, \varepsilon, \gamma_i)$ is obtained from its predecessor by an ε-transition in τ, then

$$\begin{cases} (q, \varepsilon, z, d, z) \in \tau' & \text{if no } q_i \in F \\ (q, \varepsilon, z, f, z) \in \tau' & \text{if some } q_i \in F \end{cases}$$

(6) For all $z \in \Gamma \cup \{x_0\}$, $(f, \varepsilon, z, d, z) \in \tau'$.
(7) For all $q \in Q$, $a \in A \cup \{\varepsilon\}$ and $z \in \Gamma$, if no transition of τ' starting (q, a, z) has been defined by (2) or (5), and there is a transition $(q, a, z, q', \gamma) \in \tau$, then $(q, a, z, q', \gamma) \in \tau'$.

It is easy to see that M' is deterministic. Note that M', in its initial state, always begins a computation by putting x_0 on the bottom of the stack (using (1)), and this is never erased.

Suppose M' does not always scan its entire input. Then for some $w \in A^*$,

$$(q_0', w, z_0) \xrightarrow[M']{} (q, au, z_1 \ldots z_k x_0)$$

where $a \in A$ and au is a suffix of w, $z_i \in \Gamma$, $k \geq 0$, and a is never read. That is, the computation can only be continued by use of ε-transitions. In fact, the computation of M' can be continued using an ε-transition. For if M has no transition beginning (q, ε, z_1), then either by (2) or (7) M' has a transition starting (q, a, z_1), so a can be read from the tape, a contradiction. Thus M has a transition beginning (q, ε, z_1), so either by (5) or (7), M' has a transition starting (q, ε, z_1), as claimed. Repeating this argument, the computation of M can be continued indefinitely using ε-transitions, giving a sequence

$$(q, au, z_1 \ldots z_k x_0),\ (q_1, au, \gamma_1 z_2 \ldots z_k x_0),\ (q_2, au, \gamma_2 z_2 \ldots z_k x_0), \ldots$$

Note that q and all q_i are in Q, because no ε transition begins with d, and in state f, the next configuration will be in state d, using (6). Consequently, the ε-transitions used are all transitions of M. Now eventually z_1 must be erased from the stack, that is, some $\gamma_i = \varepsilon$. Otherwise (5) applies to the sequence

$$(q, \varepsilon, z_1),\ (q_1, \varepsilon, \gamma_1),\ (q_2, \varepsilon, \gamma_2), \ldots$$

(obtained by using the same transitions used in the sequence above). The first transition used is then given by (5), so q_1 is either d or f, a contradiction.

Similarly, $z_2, \ldots, z_k$ are eventually erased, leading to a configuration (q_i, au, x_0). Then for the next move, only a transition in (3) can be used, so $q_{i+1} = d$, a contradiction. Thus M' always scans its entire input.

Finally, we need to show $L(M) = L(M')$. Suppose M accepts $w \in A^*$. There is thus a computation of M beginning with (q_0, w, z_0) which scans all of w and ends in a final state. While a non-empty suffix of w remains on the tape, the transitions used are transitions of M', by (7). This gives a computation of M', using these transitions together with an initial use of the transition in (1), starting in configuration (q_0', w, x_0). If M is in a final state just after reading all of w, then M' will be in the same state just after reading w, so M' accepts w.

Otherwise, M then uses a sequence of ε-moves until a final state is reached. Either these are transitions of M', so again M' accepts w, or (5) applies and M', just after reading w, enters state f, so accepts w. Thus $L(M) \subseteq L(M')$.

Suppose M does not accept w, and consider a computation of M starting with (q_0, w, z_0). If M halts, then (whether or not all of w has been read), we obtain, using (2) and (3), a corresponding computation of M', starting with (q_0', w, x_0), which enters state d. Since M' is deterministic, it does not accept w. (In state d, only transitions in (4) can be used, and d is not a final state.)

Otherwise, the computation of M can be continued indefinitely, giving a sequence as in (5), where, if all of w has been read, no $q_i \in F$. Up to the point where M stops reading from the tape, there is a corresponding computation of M'. Then by (5), either M' enters state d, or enters state f having read a proper prefix of w. In the latter case, M' then enters state d using an ε-transition from (6). In any case, M' does not accept w. Hence $L(M) = L(M')$.   □

We shall need the lemma just proved for the next result, which is used in Chapter 4.

Lemma A.4. *If L is a deterministic language, then $L = L(M')$ for some deterministic PDA M' which has no ε-transitions beginning with a final state.*

Proof. There is a deterministic PDA $M = (Q, F, A, \Gamma, \tau, q_0, z_0)$ such that $L = L(M)$. By Lemma A.3, we can assume M always scans its entire input. Define a new PDA $M' = (Q', F', A, \Gamma, \tau', q'_0, z_0)$ as follows:

$$Q' = Q \times \{1, 2, 3\}$$
$$F' = \{(q, 3) \mid q \in Q\}$$
$$q'_0 = \begin{cases} (q_0, 1) & \text{if } q_0 \in F \\ (q_0, 2) & \text{if } q_0 \notin F \end{cases}$$

and τ' is defined as follows.

(1) If $(q, \varepsilon, z, p, \gamma) \in \tau$, then τ' contains

$$((q, k), \varepsilon, z, (p, l), \gamma) \qquad \text{for } k = 1, 2$$

where $l = 1$ if $k = 1$ or $p \in F$, otherwise $l = 2$.

(2) If $(q, a, z, p, \gamma) \in \tau$, where $a \in A$, then τ' contains

$$((q, k), a, z, (p, l), \gamma) \qquad \text{for } k = 2, 3$$

where $l = 1$ if $p \in F$, $l = 2$ if $p \notin F$, and τ' also contains

$$((q, 1), \varepsilon, z, (q, 3), \gamma).$$

Obviously M' is deterministic and has no ε-transitions beginning with a final state. Given a computation of M starting with (q_0, w, z_0), we claim that there is a corresponding computation of M' starting with (q'_0, w, z_0), such that if the computation of M ends in state q, then the computation of M' ends in state (q, k), where $k = 1$ or 2. Further, in the computations of M and M', exactly the same word has been read from the tape.

The computation of M' is constructed by induction on the length of the computation of M. Suppose the next step of this computation uses the transition $(q, \varepsilon, z, p, \gamma)$. Then the computation of M' is continued by using the corresponding transition in (1). If the next step in the computation of M uses (q, a, z, p, γ), there are two cases. If M' is in a state $(q, 2)$, then the computation of M' is continued by using the transition

$((q,2),a,z,(p,l),\gamma)$ in (2). If M' is in state $(q,1)$, the computation is continued by using $((q,1),\varepsilon,z,(q,3),\gamma)$, followed by $((q,3),a,z,(p,l),\gamma)$.

The computation of M' so constructed will be in a state $(q,1)$ if M has entered a final state since last reading a letter from the tape, and in a state $(q,2)$ otherwise. Note that M' enters a final state when $k=1$ and M reads a letter from the tape.

Suppose $w \in A^*$ and take a letter $a \in A$ (we can assume $A \neq \emptyset$, just by adding a letter to the alphabet of M). Consider the computation of M starting with (q_0, wa, z_0). Since M is deterministic and always scans its entire input, if M accepts w then it must enter a final state after reading the last letter of w and before reading the final a. Then in the corresponding computation of M', M' will enter a final state $(q,3)$ just before reading a, so accepts w. If M does not accept w, then in between reading the last letter of w and reading a, M' remains in states of the form $(q,2)$, so does not enter a final state, hence (being deterministic) does not accept w. Thus $L(M) = L(M')$. $\quad\square$

Note. With minor modification, the argument of the previous lemma can be used to show the complement of a deterministic language is also deterministic. See [20, Theorem 12.1] or [21, Theorem 10.1].

Finally we prove a result on gsm mappings needed in Chapter 5, where the terminology is explained. The proof comes from [20, Theorem 12.3].

Theorem A.5. *The class of deterministic languages is closed under inverse deterministic gsm mappings.*

Proof. Let $S = (Q_S, F_S, A, B, \tau_S, p_0)$ be a deterministic gsm and let L be a deterministic language with alphabet B, so $L = L(M)$ for some deterministic PDA M. If there is a letter in the alphabet of M not in B, we can omit it and any transitions in which it appears. If there is a letter of B, not in the alphabet of M, we can add it to the alphabet. Thus we can assume M has alphabet B, say

$$M = (Q_M, F_M, B, Z, \tau_M, q_0, z_0).$$

By Lemma A.4, we can assume M has no ε-transitions beginning with a final state. Let r be the maximum length of a word $w \in B^*$ such that some edge in the transition diagram of S has label (a,w), for some $a \in A$. We construct a PDA M' recognising $f_S^{-1}(L)$ as follows: $M' = (Q', F', A, Z, \tau', q_0', z_0)$, where:

$$Q' = \{(q,p,w) \mid q \in Q_M, \ p \in Q_S, \ w \in B^* \text{ and } |w| \leq r\}$$
$$F' = \{(q,p,\varepsilon) \mid q \in F_M, \ p \in F_S\}$$
$$q_0' = (q_0, p_0, \varepsilon)$$

The transitions in τ' are those specified in (1)–(3) below.

(1) If τ_M contains no transition beginning q,ε,z (where $q \in Q_M$, $z \in Z$), but $(p,a,w,p_1) \in \tau_S$, then τ' contains $((q,p,\varepsilon),a,z,(q,p_1,w),z)$.

(2) If $(q,\varepsilon,z,q_1,\alpha) \in \tau_M$, then τ' contains $((q,p,w),\varepsilon,z,(q_1,p,w),\alpha)$, for all p and w with $(q,p,w) \in Q'$.

(3) If $(q,b,z,q_1,\alpha) \in \tau_M$, where $b \in B$, then τ' contains

$$((q,p,bw), \varepsilon, z, (q_1,p,w), \alpha)$$

for all p and w with $(q,p,bw) \in Q'$.

It is easily seen that M' is deterministic. Suppose M' accepts $u = a_1 \ldots a_n$. Then the transitions of type (1) which it uses correspond to transitions of S having the form (p_i, a_i, w_i, p_{i+1}) (because transitions of types (2) and (3) do not change the second coordinate of the state of M'). These transitions give a computation of S with input $a_1 \ldots a_n$ and output $w_1 \ldots w_n$. The transitions of types (2) and (3) M' uses correspond to transitions of M and give a computation of M in which $w_1 \ldots w_n$ is read from its tape, ending, say, in a state q. These transitions do not change the first coordinate of the state of M'. (The third coordinate of the states of M' represents a buffer to receive output from S, using transitions (1); after a letter is read from the buffer, using transitions (3), it is erased.) At the end of the computation M' is in state $(p_{n+1}, q, \varepsilon)$. Hence $p_{n+1} \in F_S$, so the computation of S is successful, and $a_1 \ldots a_n \in f_S^{-1}(w_1 \ldots w_n)$. Similarly, $q \in F_M$, so M accepts $w_1 \ldots w_n$, that is, $w_1 \ldots w_n \in L$. It follows that $L(M') \subseteq f_S^{-1}(L)$.

Conversely, if $a_1 \ldots a_n \in f_S^{-1}(L)$, there is a successful computation of S, with input $a_1 \ldots a_n$ and output $w = w_1 \ldots w_n \in L$, where (a_i, w_i) are the labels on the edges of the corresponding path in the transition diagram. There is a computation of M accepting w. It is left to the reader to construct a computation of M', accepting $a_1 \ldots a_n$, from those of S and M. The condition on M, that no ε-transition begins with a final state, is needed because of the possibility that $w_k = \ldots w_n = \varepsilon$ for some k. It ensures that, if this happens, M' reads $a_k \ldots a_n$ from its tape. Thus $a_1 \ldots a_n \in L(M')$, as required. $\square$

Appendix B
The Halting Problem and
Universal Turing Machines

Let X be the set of numerical Turing machines, where states are renamed so that the set of states of each machine is $\{2, \ldots, r-1\}$ for some r, and where $L = 0$, $R = 1$. We can define a mapping $gn : X \to \mathbb{N}$ as we did after Theorem 2.19 (but without first modifying the machines). Then gn is a Gödel numbering, that is, it is $1-1$ and its image is recursive (exercise). There is therefore a strictly increasing recursive bijection $f : \mathbb{N} \to gn(X)$. Putting $T_m = gn^{-1}f(m)$, we obtain an enumeration $T_0, T_1, \ldots$ of numerical TM's which is effective, in that given m, $f(m) = gn(T_m)$ is computable, and from $gn(T_m)$ one can recover the states, transitions etc. of T_m.

The general halting problem is to give a procedure to decide whether T_m, on input x (i.e. started on tape description $\underline{0}1^x$) halts or not. We shall show this is unsolvable; formally, this means that $B = \{(m,x) \mid T_m$ halts on input $x\}$ is not recursive. As in Prop. 3.7, it suffices to show that $A = \{m \mid T_m$ halts on input $m\}$ is not recursive.

Suppose A is recursive. Then $\mathbb{N} \setminus A$ is r.e., so $\chi_{p(\mathbb{N} \setminus A)}$ is recursive, hence is computed by a numerical TM T which halts on input m if and only if $\chi_{p(\mathbb{N} \setminus A)}(m)$ is defined, i.e. $m \notin A$, by Cor. 2.22. By renaming, we can assume the set of states of T is $\{2, \ldots, r-1\}$ for some r and $L = 0$, $R = 1$. Then $T = T_p$ for some p.

Then by definition of A, T_p halts on input p if and only if $p \in A$, but T_p halts on input p if and only if $p \notin A$, a contradiction. Hence A is not recursive.

Of course this is related to Prop. 3.7, but is not easy to derive directly from Prop. 3.7 because of the modifications made to the Turing machines and the use of Kleene's Normal Form Theorem.

To make further progress, we shall assume the mapping $g : \mathbb{N} \to \mathbb{N}$ defined by $g(m) = gn(T'_m)$ is recursive (to prove this is a rather complicated exercise). Here, T'_n is the modified TM defined before Lemma 2.18. Taking $n = 1$ in Theorem 2.20, the proof shows that $\varphi_{T_m,1}(x) = H(m,x)$, where $H(m,x) = F(g(m), x, \mu t(G(g(m),x,t) = 0))$, where F, G are the functions in Theorem 2.20. Now H is partial recursive, so is computed by a numerical TM, say U, by Cor. 2.22. Then U, started on $Tape(m,x)$ (i.e. $\underline{0}1^m01^x$), gives exactly the same output as T_m on input x. (They either halt with tape description $\underline{0}1^y$, where $y = \varphi_{T_m,1}(x)$ if this is defined, otherwise they do not halt.) For this reason, U is called a *universal Turing machine*. It is not clear how to construct U from what was done in Chapter 2, but there is a considerable amount of literature on universal Turing machines and their construction, and their relevance to the development of the (stored program) computer. The existence of a universal machine goes back to Turing's original papers ([38], [39].

For a discussion of the halting problem for Turing machines designed to recognise languages, see [20, §7.3].

Appendix C
Cantor's Diagonal Argument

We assume familiarity with the idea of a countable set. Recall that a set is countable if it can be put into one-to-one correspondence with a subset of $\mathbb{N}$. Equivalently, a set C is countable if either $C = \emptyset$ or there is a surjective mapping $f : \mathbb{N} \to C$. In the next theorem, $2^{\mathbb{N}}$ means the set of all subsets of $\mathbb{N}$.

Theorem C.1. *There is no surjective mapping* $\mathbb{N} \to 2^{\mathbb{N}}$, *consequently* $2^{\mathbb{N}}$ *is uncountable.*

Proof. Suppose $f : \mathbb{N} \to 2^{\mathbb{N}}$ is surjective; put $Y = \{x \in \mathbb{N} \mid x \notin f(x)\}$. Then $Y = f(z)$ for some $z \in \mathbb{N}$, and $z \in Y \Longleftrightarrow z \in f(z) \Longleftrightarrow z \notin Y$, by definition of Y, a contradiction. (The argument works for any set X in place of $\mathbb{N}$). $\square$

To interpret this in terms of characteristic functions, we can write

$$Y = \{x \in \mathbb{N} \mid \chi_{f(x)}(x) = 0\}.$$

Then $\chi_Y(x) = 1$ if and only if $\chi_{f(x)}(x) = 0$, that is, $\chi_Y(x) = 1 \dot{-} \chi_{f(x)}(x)$.

Now put $F(m,n) = \chi_{f(m)}(n)$, for $m, n \in \mathbb{N}$. Then for all $x \in \mathbb{N}$,

$$F(z,x) = 1 \dot{-} F(x,x)$$

where $Y = f(z)$, and putting $x = z$ gives a contradiction. The proof is similar to some arguments used in the course (see Props. 3.5 and 3.6) and to the proof of the Gödel Incompleteness Theorem in logic. Writing the values of F as an infinite matrix:

$$
\begin{array}{llllll}
F(0,0) & F(0,1) & F(0,2) & F(0,3) & F(0,4) & \cdots \\
F(1,0) & F(1,1) & F(1,2) & F(1,3) & \cdots \\
F(2,0) & F(2,1) & F(2,2) & \cdots \\
F(3,0) & F(3,1) & \cdots \\
F(4,0) & \cdots \\
\vdots
\end{array}
$$

row m represents the values of $\chi_{f(m)}$, so if f is surjective, each subset of $\mathbb{N}$ is represented by a row; however, $\chi_Y(x) = 1 \dot{-} F(x,x)$, so χ_Y is obtained by changing the values of F on the main diagonal, indicated by the arrows. Then Y is not represented by any row, a contradiction, since a row representing it differs from the first row in the first entry, the second row in the second entry, etc. This explains the name "diagonal argument".

It follows easily that $\mathbb{R}$ is uncountable. Let B be the set of all real numbers a whose decimal expansion has the form $a = 0.a_0 a_1 \ldots$, where every a_i is either 0 or 1. (In the case of a terminating decimal expansion, add an infinite string of zeros, so $a_1, a_2 \ldots$ is always an infinite sequence, which is uniquely determined by a). Every such number $a = 0.a_0 a_1 \ldots$ defines an element of $2^{\mathbb{N}}$, say X_a, by $\chi_{X_a}(i) = a_i$. The mapping $a \mapsto X_a$ is a bijection from B to $2^{\mathbb{N}}$, hence B is uncountable, and so is $\mathbb{R}$, as a subset of a countable set is countable.

The Russell-Zermelo Paradox

The first proof can be easily adapted to show Cantor's version of set theory is inconsistent; in this set theory, given any predicate P, there is a set $\{x \mid P(x)\}$, such that any object x belongs to the set if and only if $P(x)$ is true. Now let $y = \{x \mid x \notin x\}$. It is easy to see that $y \in y$ if and only if $y \notin y$, a contradiction.

Exercises on Appendix C

1. By explicit use of the diagonal argument, without using $2^{\mathbb{N}}$, show that the subset B of $\mathbb{R}$ is uncountable.

2. Recall from Chapter 2 that Ackermann's function is the function $A : \mathbb{N}^2 \to \mathbb{N}$ defined by

$$A(0,y) = y+1$$
$$A(x+1,0) = A(x,1)$$
$$A(x+1,y+1) = A(x,A(x+1,y))$$

It can be shown that, for any primitive recursive function $f : \mathbb{N}^n \longrightarrow \mathbb{N}$, there exists k with $f(x_1,\ldots,x_n) \leq A(k, \max\{x_1,\ldots,x_n\})$, for all $x_1,\ldots,x_n$. Use this and the diagonal argument to prove that A is not primitive recursive.

Appendix D
Solutions to Selected Exercises

Chapter 1

1. Yes, a derivation is $S, aASb, abSbSb, abSbabb, ababbabb$.

3. A transition diagram for a FSA recognising $\{(ab)^n \mid n = 0, 1, 2, \ldots\}$ is

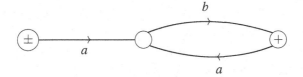

5. (ii) No. Otherwise R_L would be of finite index by Theorem 1.7, which implies $1^m 01^n 0 \; R_L \; 1^m 01^p 0$ for some $n \neq p$, where $m, n, p \geq 0$. Then

$$1^m 01^n 01^{m+n} \; R_L \; 1^m 01^p 01^{m+n}$$

a contradiction since $1^m 01^n 01^{m+n} \in L$, but $1^m 01^p 01^{m+n} \notin L$.

Chapter 2

1. (a) Let $f_n(x_1, \ldots, x_n) = \max\{x_1, \ldots, x_n\}, \underline{x} = (x_1, \ldots, x_n)$. Then

$$\begin{aligned}
f_n(\underline{x}) &= \max\{f_{n-1}(x_1, \ldots, x_{n-1}), x_n\} = f_2(f_{n-1}(x_1, \ldots, x_{n-1}), x_n) \\
&= f_2(f_{n-1}(\pi_{1n}(\underline{x}), \ldots, \pi_{n-1,n}(\underline{x})), \pi_{nn}(\underline{x}))
\end{aligned}$$

and it suffices by induction on n to show f_2 is primitive recursive. But f_2 has a definition by cases:

$$f_2(x,y) = \begin{cases} \pi_{12}(x,y) & \text{if } x \geq y \\ \pi_{22}(x,y) & \text{if } x < y \end{cases}, \text{ hence } f_2 \text{ is primitive recursive.}$$

4. Clearly $J_1^{-1} = J_1$ is primitive recursive, and J_2^{-1} is primitive recursive by Exercise 3 (b). For $n \geq 2$, if $y = J_{n+1}(x_1, \ldots, x_{n+1})$, then $y = J(x_1, J_n(x_2, \ldots, x_{n+1}))$, so $x_1 = K(y)$ and $J_n(x_2, \ldots, x_{n+1}) = L(y)$, hence $(x_2, \ldots, x_{n+1}) = (J_n^{-1} \circ L)(y)$. Thus $J_{n+1}^{-1} = (K, K_1 \circ L, \ldots, K_n \circ L)$, where $K_1, \ldots, K_n$ are the coordinate functions of J_n^{-1}. It follows by induction on n that J_n^{-1} is primitive recursive for all n. Putting $n = 2$ gives $J_3^{-1} = (K, K \circ L, L \circ L)$.

5. Suppose $\underline{a}_1, \ldots, \underline{a}_k$ are distinct elements of $\mathbb{N}^n$, and $f(\underline{a}_i) = b_i$ (where $b_i \in \mathbb{N}$) for $1 \leq i \leq k$, and $f(\underline{x})$ is undefined for $\underline{x} \notin \{\underline{a}_1, \ldots, \underline{a}_k\}$.

$$\text{Let } g(\underline{x}) = \begin{cases} b_i & \text{if } \underline{x} = \underline{a}_i, \text{ i.e } |\underline{x} - \underline{a}_i| = 0, \text{ for some } i \text{ with } 1 \leq i \leq k \\ 0 & \text{otherwise} \end{cases}$$

Then g is primitive recursive, being obtained from constant functions using a definition by cases. Now let

$$h(\underline{x}) = \mu y(|\underline{x} - \underline{a}_1| \ldots |\underline{x} - \underline{a}_k| = 0)$$

a partial recursive function. Then $h(\underline{a}_i) = 0$ for $1 \leq i \leq k$ and $h(\underline{x})$ is undefined for $\underline{x} \notin \{\underline{a}_1, \ldots, \underline{a}_k\}$. Therefore $f(\underline{x}) = g(\underline{x}) + h(\underline{x})$ is partial recursive.

7. Let H be the iterate of h, so H is primitive recursive by (the easy case of) Question 6. Then $\varphi(x, t, r) = H(x, t \dot{-} r)$, which is obtained from H and known primitive recursive functions by composition.

9. $T_1 = P_1 R^* L P_0$. The effect on the tape description is

$$u01^a 0\underline{0}1^c \xrightarrow[P_1]{} u01^a 0\underline{1}1^c \xrightarrow[R^*]{} u01^a 01^{c+1}\underline{0} \xrightarrow[L]{} u01^a 01^c\underline{1} \xrightarrow[P_0]{} u1^a 01^c\underline{0}.$$

13. Clearly $T_5 = T_3^{k-1} T_4$ will work.

Chapter 3

1. We assume $A = \{a_1, \ldots, a_n\}$ where $n > 0$ (the case $n = 0$ is easy, as noted in the text). The variables x, y, z are used below to define certain functions, and range over all elements of $\mathbb{N}$. It is a supplementary exercise to verify in detail the claims below that certain functions and predicates are primitive recursive.

(a) Let q be an integer greater than 1. If $r \in \mathbb{N}$, r can be written as

$$r = s_1 + s_2 q + \ldots + s_k q^{k-1}$$

where $0 \le s_j < q$ for $1 \le j \le k$ (by using the division algorithm and induction on r). Putting $s_j = 0$ for $j > k$, the s_j are uniquely determined. To see this, define $Q : \mathbb{N}^3 \to \mathbb{N}$ by primitive recursion:

$$Q(x,y,0) = x$$
$$Q(x,y,z+1) = \text{quo}(y, Q((x,y,z)).$$

and put $F(x,y,z) = \text{rem}(y, Q(x,y,z \dot{-} 1))$, so F is a primitive recursive function. Then the reader can check that $s_j = F(r,q,j)$ for $j \ge 1$. It follows that φ_1 is one-to-one.

Now choosing k as small as possible, k is the least integer m such that such that $r < q^m$, and $k \le r$ (by induction on r). Define a primitive recursive function $M : \mathbb{N}^2 \to \mathbb{N}$ by $M(x,y) = \mu z \le x(x < y^z)$, so $k = M(r,q)$. Now put $f(x,z) = F(x,n+1,z)$, $m(x) = M(x,n+1)$. Thus $r = \sum_{j=1}^{m(r)} f(r,j)(n+1)^{j-1}$. From the definition of φ_1

$$r \in \varphi_1(A^*) \Leftrightarrow f(r,j) > 0 \text{ for } 1 \le j \le m(r)$$

and the right-hand side is a primitive recursive predicate. Hence φ_1 is a Gödel numbering.

Also, φ_2 is one-to-one by unique factorisation into primes and

$$r \in \varphi_2(A^*) \Leftrightarrow$$

$$\left(0 < \log_{p_j}(r) \le n \text{ for } 1 \le j \le \log_2(r)\right) \wedge \left(r = 2^{\log_2(r)} \prod_{j=1}^{\log_2(r)} p_j^{\log_{p_j}(r)}\right).$$

The right-hand side is a primitive recursive predicate, hence φ_2 is a Gödel numbering.

Define $g : \mathbb{N} \to \mathbb{N}$ by $g(r) = 2^{m(r)} \prod_{j=1}^{m(r)} p_j^{f(r,j)}$. Then $g \circ \varphi_1 = \varphi_2$ and g is primitive recursive. If $X \subseteq A^*$ and $\varphi_1(X)$ is r.e. then $\varphi_2(X) = g(\varphi_1(X))$ is r.e. by Lemma 3.3(2).

Now define $g' : \mathbb{N} \to \mathbb{N}$ by $g'(r) = \sum_{j=1}^{\log_2(r)} \log_{p_j}(r)(n+1)^{j-1}$. Then g' is primitive recursive and $g' \circ \varphi_2 = \varphi_1$, so similarly $\varphi_2(X)$ r.e. implies $\varphi_1(X)$ is r.e. Thus $\varphi_1(X)$ is r.e. if and only if $\varphi_2(X)$ is r.e. Applying this to $A^* \setminus X$ and using Lemma 3.8, $\varphi_1(X)$ is recursive if and only if $\varphi_2(X)$ is recursive.

(b) As a hint, suppose $r = s_1 n + \ldots + s_k n^k$ where $1 \le s_j \le n$. Then

$$r = n((s_1 - 1) + (s_2 - 1)n + \ldots + (s_k - 1)n^{k-1}) + (n + \ldots + n^k).$$

(d) It is enough to show that $\varphi_2(B^*)$ is recursive, in view of (a) and (b), and in view of (c), we can choose the bijection $\{1,\ldots,n\} \to A$ such that $B = \{a_1, a_2, \ldots, a_s\}$, where $0 \le s \le n$. Then

$$r \in \varphi_2(B^*) \Leftrightarrow (r \in \varphi_2(A^*)) \wedge (\log_{p_j}(r) \le s \text{ for } 1 \le j \le \log_2(r))$$
$$\Leftrightarrow (r \in \varphi_2(A^*)) \wedge (\forall j \le \log_2(r)((j = 0) \vee (\log_{p_j}(r) \le s))$$

and the right-hand side is a primitive recursive predicate.

2. The construction of some of the TM's is as follows (in all cases, q_0 is the initial state).

R: has set of states $Q = \{q_0, q\}$ and transitions q_0aqaR $(0 \le a \le r-1)$.
L: defined similarly, replacing R by L in the transitions.
$\widetilde{R}$: $Q = \{q_0, q, q', h\}$, transitions

$$q_0aq_0aR \ (a \ne 0), \ q_00q0R, \ qaq_0aR \ (a \ne 0), \ q0q'0R, \ q'ahaL$$

where $0 \le a \le r-1$.

Chapter 4

1. First, we use Lemma 4.6 to convert the set of productions to

$$S \longrightarrow AA|b$$
$$A \longrightarrow aA|BBB|b$$
$$B \longrightarrow b$$

(The set $\mathscr{U}$ in the proof of Lemma 4.6 is $\{(S,S), (A,A), (B,B), (S,B), (A,B)\}$.)
Now, using the procedure in the first part of the proof of Theorem 4.7, we add a new variable C and convert the set of productions to

$$S \longrightarrow AA|b$$
$$A \longrightarrow CA|BBB|b$$
$$B \longrightarrow b$$
$$C \longrightarrow a$$

Then, using the second part of the proof, we add a new variable D and convert the productions to

$$S \longrightarrow AA|b$$
$$A \longrightarrow CA|BD|b$$
$$B \longrightarrow b$$
$$C \longrightarrow a$$
$$D \longrightarrow BB$$

giving the required grammar in Chomsky normal form.

4. (a) Hint: make use of Exercise 3

(c) If you have done parts (a) and (b), you can apply the procedure this gives to the grammar in Example (3), p.3. One possible grammar in the required form, generating $\{0^n 1^n \mid n > 0\}$, obtained by this method is

$$G = (\{A, B, S\}, \{0, 1\}, P, S)$$

where P consists of the productions

$$S \longrightarrow 0A \mid 0B$$
$$A \longrightarrow S1$$
$$B \longrightarrow 1$$

An alternative is to replace P by the set of productions consisting of

$$S \longrightarrow 0A$$
$$A \longrightarrow B1 \mid 1$$
$$B \longrightarrow 0A$$

6. A context-free grammar generating $L = \{0^m 1^m 0^n 1^n \mid m, n > 0\}$ is

$$G = (\{A, S\}, \{0, 1\}, P, S)$$

where P consists of the productions

$$S \longrightarrow AB$$
$$A \longrightarrow 0A1 \mid 01$$
$$B \longrightarrow 0B1 \mid 01$$

To show L is not deterministic, use the Pumping Lemma in the previous exercise.

Chapter 5

1. (a) From $xyx = yxy$, we obtain the consequence $x^2 yx^2 = xyxyx$, hence using the other relation, $y^7 = xyxyx = yxy^2 x$, so $y^6 = xy^2 x$. Since $y^6 = x^4$, we conclude that $x^4 = xy^2 x$, hence $x^2 = y^2$, so $y^3 = y^2$ which implies $y = 1$. Now from $xyx = yxy$ it follows that $x^2 = x$, so $x = 1$.

2. The proof, as indicated in the hint, is by induction on n. Let G_n be the group with the given presentation. It is true for $n = 2$ since G_2 is cyclic of order 2, as is S_2. (Indeed, it is true for $n = 1$ as the empty presentation presents the trivial group.) Assume $n > 2$ and $G_{n-1} \cong S_{n-1}$. By Lemma 5.2, there is a homomorphism $G_n \to S_n$ sending x_i to the transposition $(i, i+1)$ for $1 \le i \le n-1$, which is surjective as these transpositions generate S_n (an easy exercise). Hence it suffices to show $|G_n| \leq n!$.

By Lemma 5.2, there is a surjective homomorphism $G_{n-1} \to H$ sending x_i to x_i for $1 \leq i \leq n-2$, hence $|H| \leq (n-1)!$. It is therefore enough to show $(G_n : H) \leq n$. This will follow if we can show that any coset of H is in the set

$$T = \{H, Hx_{n-1}, Hx_{n-1}x_{n-2}, \ldots, Hx_{n-1}x_{n-2}\ldots x_2x_1\}$$

and to do this it suffices to show that if $Hy \in T$, then $Hyx_i^{\pm 1} \in T$ for $1 \leq i \leq n-1$. Since $x_i = x_i^{-1}$, we need to show that $Hx_{n-1}\ldots x_ix_j \in T$ for $1 \leq i \leq n$, $1 \leq j \leq n-1$. (Here $Hx_{n-1}\ldots x_i$ is to be interpreted as H when $i = n$.)

To do this, first note that $x_ix_j = x_jx_i$ if $|i - j| > 1$ and $x_{j-1}x_jx_{j-1} = x_jx_{j-1}x_j$ for $1 < j \leq n-1$.

If $i = j$, then $Hx_{n-1}\ldots x_ix_j = Hx_{n-1}\ldots x_{i+1} \in T$. If $i < j$, then

$$
\begin{aligned}
Hx_{n-1}\ldots x_ix_j &= Hx_{n-1}\ldots x_jx_{j-1}x_jx_{j-2}\ldots x_i \\
&= H(x_{n-1}\ldots x_{j+1})x_{j-1}(x_jx_{j-1}x_{j-2}\ldots x_i) \\
&= Hx_{j-1}(x_{n-1}\ldots x_i) \\
&= Hx_{n-1}\ldots x_i \in T
\end{aligned}
$$

since $j - 1 \leq n - 2$, so $x_{j-1} \in H$.

Finally, if $j < i$, there are two cases. If $j = i - 1$, then

$$Hx_{n-1}\ldots x_ix_j = Hx_{n-1}\ldots x_ix_{i-1} \in T.$$

If $j < i - 1$, then $Hx_{n-1}\ldots x_ix_j = Hx_jx_{n-1}\ldots x_i = Hx_{n-1}\ldots x_i \in T$ since $j \leq n - 2$, so $x_j \in H$.

4. Let $f : F(X) \to G$ be the extension of the inclusion mapping $X \to G$ to a homomorphism.

Assume (a) and $\alpha : X \to H$ is a mapping, where G is a group. Then α has a unique extension to a homomorphism $\tilde{\alpha} : F(X) \to H$ by Lemma 5.5. Then αf^{-1} is the unique extension of α to a homomorphism $G \to H$, hence (a) implies (b).

Assume (b). Let $\alpha : X \to F(X)$ be the inclusion map, $\beta : G \to F(X)$ the extension of α to a homomorphism. Then if $g \in G$ is represented by the non-empty reduced word u, $\beta(g) = u \neq 1$ by Lemma 5.4, so $g \neq 1$. Hence (b) implies (c).

Finally the condition on reduced words in (c) implies that $f : F(X) \to G$ has trivial kernel, and if X generates G then f is onto. Hence (c) implies (a).

6. Let X be a basis for F and let Y be a finite set of generators for F. For $y \in Y$, let u_y be a word in $(X^{\pm 1})^*$ representing y. Let X_1 be the finite subset of X consisting of all elements $x \in X$ which occur in u_y (either as x or as x^{-1}) for some $y \in Y$. Then Y is a subset of the subgroup of F generated by X_1 hence F is generated by X_1. By Question 4, no non-empty reduced word in $(X^{\pm 1})^*$ represents the identity element of F, so no non-empty reduced word in $(X_1^{\pm 1})^*$ represents the

identity element. Again by Question 4, X_1 is a finite basis for F. (It follows easily that $X = X_1$.)

8. Suppose F is free with basis X. Let F_x be the subgroup of F generated by x. Then F_x is infinite cyclic by Lemma 5.4, and the inclusion maps $F_x \rightarrow F$, for $x \in X$, extend uniquely to a homomorphism $\ast_{x \in X} F_x \rightarrow F$. This is an isomorphism by Lemma 5.4 and the normal form theorem for free products. (Alternatively, the inclusion mapping $X \rightarrow \ast_{x \in X} F_x$ extends uniquely to a homomorphism $F \rightarrow \ast_{x \in X} F_x$; show that this is the inverse map.)

For the converse, show that if F is a free product of infinite cyclic groups, then choosing a generator for each of the infinite cyclic groups gives a basis for F.

9. (b) We can take the free abelian group to be $\mathbb{Z} \times \mathbb{Z}$ with basis $\{x, y\}$, where $x = (1, 0)$ and $y = (0, 1)$. The Cayley diagram is partly drawn below.

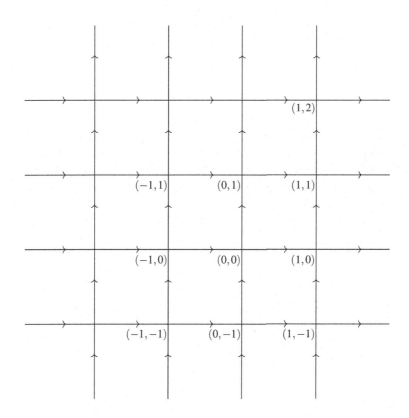

The intersections of the lines represent the vertices (the set of vertices is the set of points in the plane $\mathbb{R}^2$ with integer coordinates). The horizontal arrows have label x and the vertical ones have label y. (Usually one would use additive notation for this group, but in multiplicative notation, for example, $(1,2) = xy^2 = y^2x$.)

Removing the edges of a finite subgraph always leaves a single infinite component, so the free abelian group of rank 2 has one end.

Appendix C

1. Suppose B is countable. Then there is a surjective mapping $f : \mathbb{N} \to B$. Writing b_{n+1} for $f(n)$, we can write B in a list $B = \{b_1, b_2, \ldots\}$. By definition, we can write

$$b_1 = 0.a_{11}a_{12}a_{13}\ldots$$
$$b_2 = 0.a_{21}a_{22}a_{23}\ldots$$
$$\vdots \quad \vdots$$
$$b_i = 0.a_{i1}\,a_{i2}\,a_{i3}\ldots$$
$$\vdots \quad \vdots$$

where a_{ij} is either 0 or 1, for all integers $i,\ j \geq 1$. Define $a_i = 1 - a_{ii}$ for $i \geq 1$, then put $b = 0.a_1a_2a_3\ldots$, an element of B since a_i is either 0 or 1 for all i. Therefore $b = b_i$ for some i, which is impossible as the decimal expansions of b and b_i differ in the ith position ($a_i \neq a_{ii}$), a contradiction. Hence B is uncountable.

References

1. J.W. Backus, "The syntax and semantics of the proposed international algebraic language of the Zürich ACM-GAMM conference". In: *Proc. Int. Conf. Inf. Process., Paris 15-20 June 1959*, pp 125–132. UNESCO 1960.
2. M.R. Bridson and R.H. Gilman, "Formal language theory and the geometry of 3-manifolds", *Comment. Math. Helv.***71** (1996), 525–555.
3. D.E. Cohen, *Groups of cohomological dimension one*, Lecture Notes in Mathematics **245**. Berlin, Heidelberg, New York: Springer-Verlag 1972.
4. D.E. Cohen, *Computability and logic*. Chichester: Ellis Horwood; New York etc.: Wiley (Halsted Press) 1987.
5. D.E. Cohen, *Combinatorial group theory: a topological approach*, London Math. Soc. Student Texts **14**. Cambridge: University Press 1989.
6. M.J. Dunwoody, "The accessibility of finitely presented groups", *Invent. Math.* **81** (1985), 449–457.
7. D.B.A. Epstein, J.W. Cannon, D.F. Holt, S.V.F. Levy, M.S. Paterson and W.P. Thurston, *Word processing in groups*. Boston etc.: Jones and Bartlett 1992.
8. S.M. Gersten and H. Short, "Small cancellation theory and automatic groups. II", *Invent. Math.* **105** (1991), 641–662.
9. R.H. Gilman,"Formal languages and infinite groups". In: *Geometric and computational perspectives on infinite groups. Proceedings of a joint DIMACS/Geometry Center workshop, January 3-14, 1994 at the University of Minnesota, Minneapolis, MN, USA and March 17-20, 1994 at DIMACS, Princeton, NJ, USAG* (eds G. Baumslag et al), *DIMACS Ser. Discrete Math. Theoret. Comput. Sci.* **25**, pp. 27–51. Providence, RI: Amer. Math. Soc. 1996.
10. R. Gregorac, "On generalized free products of finite extensions of free groups", *J. London. Math. Soc.* **41** (1966), 662–666.
11. R.H. Haring-Smith, "Groups and simple languages", *Trans. Amer. Math. Soc.***279** (1983), 337–356.
12. M.A. Harrison, *Introduction to formal language theory*. Reading, Mass. etc.: Addison-Wesley 1978.
13. T. Herbst, "On a subclass of context-free groups", *Theor. Inform. Appl.* **25** (1991), 255–272.
14. T. Herbst and R.M. Thomas, "Group presentations, formal languages and characterizations of one-counter groups", *Theoret. Comput. Sci.* **112** (1993), 187–213.
15. G. Higman, B.H. Neumann and H. Neumann, "Embedding theorems for groups", *J. London. Math. Soc.* **24** (1950), 247–254.
16. D.F. Holt and S.E. Rees, "Solving the word problem in real time", *J. London. Math. Soc.* **63** (2001), 623–639.
17. D.F. Holt, S.E. Rees, C.E. Röver and R.M. Thomas, "Groups with context-free co-word problem", *J. London Math. Soc. (2)* **71** (2005), 643–657.

18. D.F. Holt and C.E. Röver, "On real-time word problems", *J. London. Math. Soc.* **67** (2003), 289–301.
19. D.F. Holt and C.E. Röver, "Groups with indexed co-word problem", *Internat. J. Algebra Comput.* **16** (2006), 985–1014.
20. J.E. Hopcroft and J.D. Ullman, *Formal languages and their relation to automata.* Reading, Mass. etc.: Addison-Wesley 1969.
21. J.E. Hopcroft and J.D. Ullman, *Introduction to automata theory, languages and computation.* Reading, Mass. etc.: Addison-Wesley 1979.
22. J.E. Hopcroft, J.D. Ullman and R. Motwani, *Introduction to automata theory, languages, and computation.* Reading, Mass.: Addison-Wesley 2001.
23. A. Karrass and D. Solitar, "The subgroups of a free product of two groups with an amalgamated subgroup", *Trans. Amer. Math. Soc.* **150** (1970), 227–255.
24. A. Karrass, A. Pietrowski and D. Solitar, "Finite and infinite cyclic extensions of free groups", *J. Austral. Math. Soc.* **16** (1973), 458–466.
25. R.C. Lyndon and P.E. Schupp, *Combinatorial group theory*, Ergebnisse der Math. **89**. Berlin, Heidelberg, New York: Springer 1977.
26. W. Magnus, *Noneuclidean tesselations and their groups.* New York, London: Academic Press 1974.
27. D.E. Muller and P.E. Schupp, "Groups, the theory of ends, and context-free languages", *J. Comput. System Sci.* **26** (1983), 295–310.
28. P. Naur et al., "Report on the algorithmic language Algol 60", *Comm. ACM* **3** (1960), 299–314 (ibid. **6** (1963), 1–17).
29. R. Péter, *Recursive functions. 3rd revised ed.* New York, London: Academic Press; Publishing House of the Hungarian Academy of Sciences 1967.
30. D.W. Parkes and R.M. Thomas, "Groups with context-free reduced word problem", *Comm. Algebra* **30** (2002), 3143–3156.
31. J-E. Pin, "Finite semigroups and recognizable languages: An introduction". In: *Semigroups, formal languages and groups, Proceedings of the NATO Advanced Study Institute, York, UK, August 7–21, 1993* (ed. J. Fountain), pp 1–32. Dordrecht: Kluwer Academic Publishers 1995.
32. J.J. Rotman, *An introduction to the theory of groups*, Graduate Texts in Mathematics **148**. New York: Springer-Verlag 1995.
33. G. Rozenberg and A. Salomaa (eds.), *Handbook of formal languages* Vols. 1–3. Berlin: Springer-Verlag 1997.
34. A. Salomaa, *Formal languages* ACM Monograph Series. New York, London: Academic Press [Harcourt Brace Jovanovich] 1973.
35. J.-P. Serre, *Trees.* New York: Springer 1980.
36. M. Shapiro, "A note on context-sensitive languages and word problems", *Internat. J. Algebra Comput.* **4** (1994), 493–497.
37. H. Short, "An introduction to automatic groups". In: *Semigroups, formal languages and groups, Proceedings of the NATO Advanced Study Institute, York, UK, August 7–21, 1993* (ed. J. Fountain), pp 233–253. Dordrecht: Kluwer Academic Publishers 1995.
38. A.M. Turing, "On Computable Numbers, with an Application to the Entscheidungsproblem", *Proc. London Math. Soc. Ser. (2)***42** (1937) 230–265. (Reprinted in *The Undecidable* (Ed. M. David). Hewlett, NY: Raven Press, 1965.)
39. A.M. Turing, "Correction to: On Computable Numbers, with an Application to the Entscheidungsproblem", *Proc. London Math. Soc. Ser. (2)***43** (1938) 544–546.
40. B.L. van der Waerden,"Free products of groups", *Amer. J. Math.* **70** (1948), 527–528.

Index